WHAT WOOD IS THAT?

A Manual of Wood Identification

with
40 actual
wood samples
and
79 illustrations
in the text

Herbert L. Edlin
B.Sc. Forestry

VIKING

VIKING
Published by the Penguin Group
Penguin Putnam Inc., 375 Hudson Street,
New York, New York 10014, U.S.A.
Penguin Books Ltd, 27 Wrights Lane,
London W8 5TZ, England
Penguin Books Australia Ltd, Ringwood,
Victoria, Australia
Penguin Books Canada Ltd, 10 Alcorn Avenue,
Toronto, Ontario, Canada M4V 3B2
Penguin Books (N.Z.) Ltd, 182–190 Wairau Road,
Auckland 10, New Zealand

Penguin Books Ltd, Registered Offices:
Harmondsworth, Middlesex, England

First published in 1969 by Viking Penguin Inc.
Published in Great Britain and the Commonwealth by
Stobart Davies Ltd, Priory House, 2 Priory Street,
Hertford SG14 1RN

17   16   15   14   13   12   11   10   9   8

ISBN 0-670-75907-4
Library of Congress Catalog Card Number: 69–15933

Printed in Great Britain by BPC Wheatons Ltd., Exeter
Set in Monotype 10/12 pt Garamond

Contents

Sequence of Trees and Timbers Described

Introduction

Wood is the most variable and adaptable raw material available to man. We use it for a wider range of purposes than any other substance. Modern technology has brought many possible alternatives, from plastics and metals to concrete and earthenware, yet timber still holds its own. This is due to a unique combination of qualities which it derives from its natural growth within the living tree.

Every piece of wood you handle once served to support a tree's leaves and branches, and to carry sap up to nourish them. It must therefore be basically strong, yet it is light. For once it has been seasoned its cells are full of air. Because it is made up of plant tissues, it is soft in comparison with stone or iron, and this makes it much easier to work, either with simple hand tools or by machinery. Everyone can tackle simple jobs needing wood, with cheap and familiar tools, but mineral substances need the training and special equipment of a technician.

Wood, though hard and tough enough for everyday wear, is somehow sympathetic to our needs. It does not present the sharp or harsh edges of metal or concrete, and – barring the odd splinter – you are unlikely to hurt yourself on any piece of wood. Its open meshwork of air-filled cells makes it a poor conductor of heat. Wooden objects are never cold to the touch, and thin wooden walls keep out the cold much better than thicker structures of brick, concrete or stone.

Timber is also plentiful and cheap. Forests grow in most regions of the world except for the prairies, the highest mountains, and the great hot or ice-cold deserts. Wood can be harvested and shaped at low cost, and is in fact transported economically to every civilized country, regardless of where it first grew. Since it grows naturally, though slowly, it is a renewable resource. Even before the foresters felled the oak that now makes your kitchen table, they had planned to replace it with acorns to yield the next oak crop – due for harvesting one hundred years ahead.

Every kind of tree produces a different fine structure of cells making up its timber, but this is apparent only under the microscope. It results in surface patterns, colours and lustres that the eye can readily detect, and in qualities of texture, weight and hardness that are easily recognized. Though each pattern is constant for that kind of tree, it is expressed in different ways in separate individual trees, and even within different parts of the same tree. The method of cutting and finishing each wooden object will also give, quite literally, a particular slant on its intimate structure.

Though all wood is basically similiar, every fragment shows the variability found in all natural materials. No two pieces of wood in the world are exactly alike. This gives timber its unique fascination, for no other common durable substance shows patterns and properties peculiar to each piece. At the same time it makes identification tricky. You have to learn to isolate those features that are characteristic of a certain timber, from others that many kinds of woods can share.

The first aim of this book is to enable you to single out the essential properties of forty timbers in regular use today, so that you can recognize each after brief inspection. Its key feature is the inclusion of forty actual specimens that show the essential points better than any printed illustration. No two copies of the book itself can ever be identical, for there will inevitably be subtle differences among the forty specimens included, though all are fair examples of their kinds.

The second object is to show how these woods, brought together from a score of countries scattered over the five continents, have grown in their native forests. The properties of each, which ensured their selection, are described, and so are the processes that are used to transform them from standing trees and huge round logs to familiar furniture or household goods.

Every piece of wood brings with it a hint of the far forest where it grew, and a touch of the romance of trade and technical skills that have transformed a living tree for you to use, admire and treasure.

As you master the peculiarities of the grain and fine structure, you will find a growing interest in every wooden object you see around you. Here is a road to fresh horizons, for the greatest expert never knows 'all about wood'. There are always challenging new pathways to explore in this exacting world of timber, just as there are always new tracks through the forest itself.

PART I: THE WOOD FROM THE TREES

1 Man Masters Timber

From the very earliest times man has used wood for tools and weapons, buildings, furniture, or fuel for his fires. Archaeologists date primitive cultures by the material that men used, as much for the cutting and shaping of timber as for the chase or for war. In Europe the polished stone or chipped flint axe-head marks the Stone Age, the cast bronze axe the Bronze Age, and the iron axe the Iron Age, in which we still live today. But although the materials for the axe-head have changed, the wood preferred for the handle has been the same over thousands of years – tough and supple ash. Men quickly learnt that the wood of one tree served their needs better than that from another. They linked the shape of leaf and the texture of bark with the strength of the timber, and gave each tree a distinctive name.

Modern technology has confirmed these prehistoric discoveries. Every piece of ash wood, for example, is threaded through with large pores set in distinct rings, and these alternate with rings of quite different substance – strong, hard, heavy summerwood. This remarkable pattern makes ash very strong and yet elastic, able to stand up to repeated hard shocks when used as the handle of an axe or a hammer. In contrast, beech, which has an even structure all through, is brittle and apt to split when subjected to any sudden strain.

Valuable qualities were quickly recognized in other timbers, and each received a name. Elm was hard to split, so it made good chair-seats; the heartwood of oak would endure indefinitely, even in contact with damp ground, so it was chosen for fence posts and building timbers; while walnut was very stable and made the best gun-stocks. The first settlers in New England were quick to make comparisons between the native American trees and those of Europe. Many woods proved similar, but others were new and strange. Western red cedar, for example, has no equal anywhere in Europe for lightness combined with strength and durability, and has therefore remained a prized timber, for both home use and export, ever since its discovery. As travellers explored the tropics, they found that there, too, grew many woods with exceptional qualities already recognized by the Asiatics and the Amerindians. Teak and mahogany were soon exported to northern lands where they could not be matched for great structural strength or attractive appearance and workability.

Until quite recent times man was entirely dependent on hand tools to fell trees and work up their timber. Machines that would saw up logs, and plane and finish their timber in factories, did not come into general use until the nineteenth century. Machines that can go out into the forest to fell and haul in lumber are nearly all twentieth-century inventions. Even now horses, oxen, buffaloes and elephants still haul much of the world's wood from stump to highway. To see how man has mastered timber we must begin with the simplest hand tools, axe and saw, and trace their development. As their power increased, people were able to make effective use of ever larger and stronger timbers.

Axes

Man is a tool-using animal. His progress towards civilization first began when he grasped sticks and stones and discovered that these greatly increased the effectiveness of his unaided hands. One day some inventive pioneer wedged a sharp-edged stone in the cleft of a stick and discovered that he had a better tool and weapon than either a stick or a stone alone. Very soon men discovered that certain kinds of stone were better than others for the blade just as certain trees gave tougher handles than others. At Grimes Graves near Thetford in East Anglia you can still see caves where men dug hard flints out of the soft chalk and chipped them to gain a razor-sharp edge, and at Penmaenmawr in North Wales early toolmakers shaped and polished hard granite rock to the same end. From such centres good axe-heads were exported for hundreds of miles along prehistoric trade routes.

But one problem always faced the makers of stone axes and was never entirely overcome. Stone heads tended to split their handles and several clever dodges were used to overcome this. Knotty wood was used, or the handle was bound with leather thongs, or else the stone was fixed in a socket of bone. In some remote parts of the world tribesmen still use such stone axes for all their wood-cutting even today.

Once the early people of the Near East learnt to win metals from their ores, they quickly developed bronze axes. These were small, elegant tools, for the copper and tin, which were mixed to make bronze, were scarce and precious. The heads were hollow within, so that they could be fixed on to a branch stub at the side of the handle, and this device overcame the problems of splitting. The fuel used to smelt the bronze was charcoal, and men needed axes to fell the wood for charcoal-burning – in order to make more axes!

About the year 1000 BC men found that a much harder metal could be won, by using greater heat, from more plentiful ores. This was iron, and with its later development into steel it gave us the axes we still use today. With iron our forebears cleared land for crops and won all their building timber and fuel wood. The iron axe-head, with its 'eye' to take the shaft, is still the major cutting tool in most forested countries, though the recent development of the power-saw in North America and Europe is beginning to push it into the museums!

You can use an axe in three ways, according to the grain of the timber. It will cut *across* the trunk or log, split it radially in the craft of *cleaving*, or shape it to a square outline in the craft of *hewing*. Early settlers in New England used all three methods to build their homesteads – for despite the log-cabin tradition these were made of oak. Log-cabins came in later, with Scandinavian immigrants.

First they felled the plentiful oaks, using the two traditional strokes – the down stroke that splits out a segment of wood at the base of the trunk and the lower, level stroke that cuts it out as a chip. Small trees were felled with the axe alone. For larger ones the felling was finished with a saw. Branches were trimmed away with the axe and the logs were then cross-cut to the desired lengths with the saw. Then the hewing began.

Using a flat-sided hewing-axe the woodcutter worked down four sides of the round log, each in turn, splitting off the outer layers to give a squared beam. In this process all bark and most perishable sapwood were cut away to leave a squared baulk of heartwood that made a strong and durable upright post or cross-beam. Pioneers were not worried about dead-straight edges or dead-flat sides. They did not have to consider economy of timber but only economy of effort, and hewing proved far faster than sawing.

Clapboard for the sides of houses and barns, and shingles for the roof, were made by cleaving. A log was first struck through the centre with the axe. This made the first cleft, which was enlarged, if it proved stubborn, with wedge and mallet, until two

9

half-cylinders fell apart. The axe was then used to cleave out narrow pieces of wood from each segment, working always along the 'true quarter' or radius of the log, where the wood splits easily along the rays of its grain. Flat, thin sheets of wood were quickly produced in this way. They proved exceptionally strong and durable, since the minimum of cutting had been done through the timber's cells.

Two special tools derived from the axe were also used. One of these, called the *froe* or *frow*, was employed to split off roofing shingles. The other, called the *adze*, was used in hewing to get a smooth surface. Very old timbers may be known as such by the marks of the adze – quite different in character from the straighter surface left by the saw or the plane.

Cleaving was also used to cut durable fence posts and rails from round logs. A skilled man could make his house, barn, fences and furniture from standing trees using little more than an axe and his own muscle, and this versatility proved of great value when settlers moved westwards.

Wood structure revealed by the axe

Whenever an axe-cut is made across an oak tree or log, a pattern of circles or rings is revealed. Although these rings have no constant width, they increase in number with increasing thickness of stem and greater age of the tree. It was soon appreciated that each ring represented a year in the life of the tree or branch concerned, and so they were called *annual rings* or *growth rings* (see Fig. 18). People did not yet know how they arose, for wood formation remained a mystery until diligent botanists applied their microscopes. Craftsmen soon realized that the rings gave a useful clue to the wood's working properties. Close-ringed oak, for example, was 'milder' and easier to work than wide-ringed stuff, though less resistant to heavy strains. Most timbers showed this clear ring pattern. Colour and texture on the 'end-grain', as it was called, helped people to name them. Certain timbers had rings that could scarcely be seen, and again this feature was constant for each sort of tree. Some timbers, though not all, showed narrow rays spreading out from the centre (Fig. 20).

The surface of hewn timber shows another pattern, for the axe or adze cuts through the sides of the cylinders formed by the rings. Hence they appear as narrow bands making irregular flame-like shapes, often a series of ellipses one within another. This again is clear in some timbers, hard to see in others (Fig. 21).

Cleaving disclosed another pattern, for the cleaving-axe or froe always follows the path of the narrow rays that run out from the heart of the tree towards its circumference. Bright patches of ray tissue appear in consequence on many, though not all, timbers (Fig. 20).

The significance of all this for those who seek to name timbers will be explained in Chapter 3.

Saws

The very first saws used by prehistoric man were made by chipping jagged edges on pieces of hard flint, which then were mounted on handles of wood or bone. These saws were very small and were used only for the shaping of small objects of wood, bone, horn, soft stone or soft metals, to make ornaments. Saws large enough to fell trees or work up timber could not be made until harder metals were smelted. The early Egyptians used saws of copper, and the Romans had iron saws not unlike some still used today. Nowadays, all saws are made of tempered steel in a wide range of patterns to suit their work. Many kinds have been mechanized, but for centuries every blade had to be pushed or pulled by hand.

All hand-saws, and most mechanized ones, consist of a metal blade carrying sharp teeth on one edge. These teeth are carefully designed, and are sharpened so that each removes a fragment of wood in turn. They are also given a calculated amount of *set*, which means that successive teeth are bent slightly towards opposite sides. Hence the saw cuts a gap or *kerf* that is wider than the main blade, and this prevents it from jamming in the cut. This design enables the saw to open up a narrow gap through the timber, with great accuracy and minimum waste of wood, regardless of the direction of the grain. The saw is a precision tool, in contrast to the axe, which can only cut or shape wood crudely, since it depends on impact.

Though saws can ignore the grain, most kinds are designed to work either across it or with it. Hand cross-cut saws of various patterns were used, in conjunction with the axe, as the main tree-felling tool until about 1930. They were also the standard implement for cross-cutting logs to required lengths out in the woods. During the last forty years they have steadily lost ground to the petrol-driven chain saw, which works on an entirely different principle. Instead of a flat blade being drawn to and fro to rasp sharp teeth over the surface of the wood, an endless chain is propelled round a fixed guide-blade. This chain carries large sharp teeth that tear out large chips of wood. This cut is crude, but it·is quickly and cheaply made, with little effort by the operator.

For centuries all lengthwise sawing of planks and beams could only be done by hand-power, using a huge pit-saw. The log was mounted over a deep brick-lined pit, and two men drew the great blade up and down for hours on end. One man stood below it and was showered with sawdust. The other stood on top and had the heavier tasks of lifting the heavy blade and guiding its course through the wood. Eventually, in about 1420, ingenious Germans at Breslau in Silesia began to apply the power of water-mills to the movement of the heavy saw, but its blade still went up and down as before. Reciprocating saws, working either as single blades or in 'gangs' that cut out many planks at one time, have now been fully mechanized. They still operate in many timber-producing countries, using steam, oil or electric power.

About 1781 Walter Taylor, a sawmiller at Southampton in England, began to saw up logs with circular blades, using the power of a water-wheel on the River Itchen. He cut sharp teeth round the edge of a round metal plate, which he revolved on a spindle, and found it quickly cut through logs pushed against it. Few inventions have been developed so quickly. Within a few years, hundreds of circular sawmills were at work wherever there was water power to drive them, and the coming of steam power, diesel engines and electricity have further extended their use.

The sharpening and tempering of the steel blade is highly skilled work, and this made it hard for sawmillers to operate in remote districts, far from 'saw doctors'. In 1824, however, R. Eastman of Brunswick, Maine, invented and patented the inserted-tooth circular saw, which enabled sawyers to replace worn teeth by new sharp ones, instead of resharpening the whole blade.

In 1808 William Newberry of London patented a saw with teeth set on one edge of an endless metal band, revolving round two wheels; but he was unable to construct a satisfactory commercial model because the steel available at that date would not stand the strain of use. Practical bandsaws were first made by Périn of Paris about 1855. Some operate with the band moving vertically, others have it set horizontally. They can cut larger logs than circular saws, and do it faster and more accurately, but are less suited to the rougher work of small woodland mills.

Once lengthwise sawing had been mechanized, mills were set up wherever there were forests to feed them, and transport to markets was at hand. The old crafts of hewing and cleaving were largely abandoned, persisting only in remote nooks of the woods. The Industrial Revolution in North America and Europe demanded vast quantities of accurately sawn lumber for buildings and furniture, railway carriages, shipbuilding, and packing cases. The new sawmills supplied it. Nearly every piece of squared timber you handle today has come through a mechanized sawmill. The great advantages of the saw over the axe are that it can be given great external power, while it can also be adapted to all sorts of hardwood and softwood, and to any direction or peculiarity of the timber's grain.

Structure revealed by the saw: directional sawing

If a log is cut through lengthwise to give planks of constant width, each cut will expose a different view of the hidden structure of the wood. Any saw-cut that passes right through the heart of the tree will reveal the ray figure, just as though the log had been cleft with an axe (Figs 19 and 20). Cuts farther out will show 'slash grain', because, like the blade of a hewing axe or a froe, it strikes through the sides of the annual rings to expose them as irregular ellipses (Fig. 21).

The great bulk of the world's lumber is sawn up regardless of these features of surface figure. It is cut on the 'through-and-through' plan, just as it comes to the saw. For this reason, few pieces of wood ever show the same figure on opposite sides. This adds to the charm of wood as a constructional material, but also brings in a variety of appearance that may be unwanted by the designer of fine furniture, joinery or even flooring. Also, each piece of timber will vary a little in strength and wearing properties, which depend on the lie of the grain.

Sawyers soon found that if they adjusted the direction of the log before each saw-cut was made, they could produce planks with fairly constant patterns and physical properties. A cut through the centre of an oak log gives two surfaces that expose attractive silver grain; one on each semicircular segment of the log. So two thin planks can be cut from each log that are, in the timber trade's term, 'quarter-sawn'. Because all the annual rings can be seen 'on edge', such lumber is called 'edge-grained'. It is consistent in pattern and properties, and harder-wearing and more stable than ordinary 'run-of-the-mill' timber. Unfortunately, it is much more costly, because once the first two surfaces have been sawn from the log, it is impossible to cut others that are so wide. Even the narrower edge-grained planks that are cut later create a great deal of waste.

It is also possible to saw planks that show an attractive 'slash-grain' figure, but the need to make the saw-cuts well out from the heart of the tree trunk means that such planks must be narrow, and again waste occurs.

Wood sawn directionally to give consistent surface patterns can be found in high-quality antique furniture and joinery. Some is still used today in the less costly, though still highly attractive, hardwoods such as elm and oak. High-class flooring of pine will often be found to show edge-grain, because of its superior stability and wearing properties, as well as improved appearance. But, except for a few specialized purposes, directionally sawn timber has nowadays been supplanted by surface *veneers*, discussed in the next section.

Most timber that needs identification will probably have been 'converted' from the round log by sawing. It will often help you to understand structure if you stop to think *how* it was sawn. You may, for example, get a plank of edge-grained, quarter-sawn wood showing silver grain, probably much clearer on one surface than on the other. Alternatively, a plank sawn by the through-and-through method will show a distinctly different appearance on one side, compared with that on the other. The structural rings and rays will be revealed at different angles, giving good clues to their character.

Veneers

Cabinet-makers who used scarce and valuable woods of beautiful figure and colour soon sought some means to make a little material go a long way. The obvious solution was to saw a plank of dark ebony or glowing, coppery-red mahogany into thinner slices and to glue each neatly over the surface of some cheaper timber. Wood responds well to this treatment, and even when fixed with the simple fish-glues used centuries ago it remains firm and stable even today.

The earliest veneers were all hand-sawn and can be recognized as such by their thickness. When the method was applied to the directional cutting of an oak log, it was possible to cut many more surfaces that were exposed 'on the true quarter', and to reveal their bright silver grain. But sawing was tedious, and when many thin plates were cut from a costly log, half of its substance vanished in sawdust!

Sliced veneers

The next advance was the cutting of these valuable sheets of wood by a slicing process, to eliminate all loss of wood in sawdust and also speed up the job. The log is first sawn into segments convenient to handle. Each segment is then steamed to make it both softer and more supple, so that thin sheets are unlikely to crack when split off the main log. The segments, or 'flitches' as they are termed, are then fixed in a veneer-cutting machine that moves against them a very sharp, strong knife. This cuts off a thin slice from the wood surface, and the process is repeated, with adjustments of the machine, until the whole flitch has been cut into a series of sliced veneers.

The machine can be set to cut in any direction relative to the grain of the wood. The commonest direction is radial, which reveals, for example, silver grain in oak and the brightly alternating stripes of sapele and afrormosia.

Crown-cut veneers are sliced from flitches with their length set parallel to the knife's direction, but arranged to be cut clear of the log's centre – i.e. tangentially, not radially. The surfaces so exposed show annual rings crossing them obliquely with a great variety of curvature, grain and pattern. They have acquired this name because some of the finest figure arises from cuts made through the crown of a halved tree-trunk, which give a fascinating variety of grain direction.

Butt veneers often bring in the root-spurs at the very base of the tree-trunk, again adding to the fresh range of unusual patterns. Walnut is often cut this way.

Burr veneers are sliced through the burrs or swellings that arise, through some freak of nature, on the sides of certain trees. These burrs may be due to insect attack, or just result from an inherent tendency to multiple bud formation. Burr veneers usually disclose a pattern of buried branches or minute knots, such as those seen in Bird's-eye maple (see p. 104).

Rotary-cut veneers

Another kind of veneer is produced, on an enormous scale, by the related, but quite distinct, process called 'rotary cutting' (Illustrated in Fig. 22). This is an entirely modern process, made possible only by the development of powerful machinery. Briefly, a large log is first steamed to make it soft and supple, and then fixed in a huge lathe which rotates it. As it turns, a long, sharp knife is pressed against it, so that a thin sheet of wood is peeled from its circumference. After each revolution this knife is automatically moved a bit closer to the heart of the tree, so that the peeling motion can continue. This results in a sheet of veneer that can be many yards long, and as wide as the log is long. In practice this thin sheet is cut into sections of convenient length as it emerges from the veneer lathe.

From the nature of the process, rotary-cut veneers can expose only the slash grain of the wood; they can never show the ray figure of a radial section. In many woods, however, this slash grain is remarkably attractive, and rotary-cut veneers have the great practical advantage of greater size; they are not limited by the thickness of a log. The process is used for some of the most costly and strikingly coloured woods, such as bubinga, which reveal brilliant patterns when so cut, and also for more homely timbers like birch, Douglas fir, and ponderosa pine. If the logs concerned happen to hold decorative figure, this is at once apparent and effective, despite the low cost of the raw material.

Rotary cutting is also the basis of the plywood industries. Plywood is made by gluing sheets of veneer together, with the grain of each sheet running at right angles to that of its neighbour, so that the resulting plywood, though thin, is strong and stable in all directions, and very hard to split.

A feature of rotary-cut veneer is that any pattern of grain or knots within a log is repeated at every turn of the log in the lathe. There is always a slight variation in appearance, and the distance between repetitions slowly lessens, but this variability adds to the effectiveness of such veneer when used for large surfaces, such as wall decoration. Veneers of all kinds are frequently 'matched' by the skilful selection of related sheets showing a common pattern of grain and colour.

The great value of the veneer process as a whole is that it enables a large number of people to own, see and enjoy that remarkable beauty of figure that can lie hidden within a single, choice log of wood. It is also a sound economic process because the cost of finding and processing such a log is spread in small amounts over many purchasers.

Planing

Sawn wood inevitably has a rough surface, because the action of the saw-teeth breaks the fibres of the timber. Everyone is familiar with the joiner's hand plane, which drives a sharp, flat blade over the surface of sawn wood to make a simple straight cut and leave a smooth surface. In industrial processing this is replaced by powerful planing machines with revolving cutters. Planks are passed rapidly through these machines, which cut smooth faces on as many sides as required. It is this mechanization of slow processes developed by hand that makes timber so cheap in bulk, and enables it to compete commercially with other materials.

Planed surfaces are of great help to anyone who has to identify timbers, because they show up internal structure so clearly. If you have any doubts about the name of a piece of rough wood, a few strokes of a hand plane will often resolve them.

Early carpenters and joiners used hand planes to cut tongues, grooves and similar jointing devices on the edges of their planks. This work, too, has all been mechanized and speeded up, and machines will also bore mortises or shape tenons.

Carving, shaping and moulding

So far we have looked at processes for preparing straight timber. How are the curved surfaces that once needed hours of patient hand carving, with the chisel, achieved in modern commercial practice? Part of the cutting can be done by suitable fine bandsaws. The operator marks a curved outline on a flat, squared piece of wood and then manipulates it against the narrow, speeding blade of the saw to cut out the bent shape. This leaves the rough edges inevitable with sawing, so the work is completed on a remarkable machine called a 'spindle moulder'. This has an upright spindle projecting through a flat working surface, and carrying a sharp cutting-blade. It revolves at high speed and enables the worker to obtain smooth, curved edges to his wood by pressing the timber lightly against it.

Wood-turning

Many familiar wooden objects such as bowls and handles have a round cross-section, and these are always made by the process of turning. A suitably shaped piece of wood is fixed in a lathe, which revolves it rapidly. As it turns, a sharp chisel is held against it, by a man or a machine, so that a thin layer is skimmed off and the wood gradually assumes a perfectly circular outline. Deeper cutting will change a flat cylinder into a hollow bowl. Turning was originally a simple hand craft, with the power supplied by the turner's foot, pressing on a treadle. Like other forms of wood-working, it has been converted to cheap mass production on automatic lathes. A few craftsmen turners survive, and their work is highly prized. A wealth of turned woodware can be found in antique shops, some of it the outcome of hand craftsmanship, some of power-lathes; only an expert can say which.

Because wood grows in round tree trunks and branches, people rather expect round turned objects to show the same arrangement of structure as a growing stem. In other words, the heart of a tree should be found at the centre of a round bowl or handle. But wood cut in this way is very apt to warp and split, so that method is seldom used except for cheap round or half-round goods such as broom-heads.

Round or shaped tool-handles are cut with the grain of the timber running lengthwise, but always from wood taken from a position well out from the heart of the tree. Bowls and similar objects are cut, remarkably, with the grain of the wood running across the hollow of the bowl. If you wish to see the annual rings as circles, you must look at one side of the bowl, not at its centre.

Bending

So far all the wood we have looked at, though worked up in many ingenious ways, has remained straight. There is another wood-working process that can give it an entirely new shape, with results that puzzle those who set out to name specimens. If wood is treated with steam, it becomes, for the time being, plastic, and can be bent without breaking. After it has cooled and dried in its bent form, it 'sets' that way, and will never regain its original straight direction. Bending is often applied to beech in the making of 'bentwood' furniture, and also to ash when shaping handles, hockey-sticks or other curved implements.

U-bends are quite common, and if required, wood can be formed into a complete circle.

Laminated timber

A 'lamina' is a layer of substance, and if several layers are built up together you get a laminated material. This process is often applied to wood because it gives a composite material that may be stronger, tougher or more attractive than one simple piece. It also makes possible the production of larger units. Plywood, discussed earlier, is the best-known example, but there are many similar substances such as blockboard and laminboard. Very large and strong beams and trusses are nowadays constructed by laminating many small lengths of wood together.

It is quite a common practice to use timbers of different kinds when constructing laminated wood (Fig. 1). This can prove

Fig. 1 Cross-section through a laminboard. The middle layers are composed of thick rotary-cut veneers, which are covered on both sides with sheets of ordinary plywood-type veneer

confusing to those who seek to name the resulting 'timber'. Where doubt arises, ask yourself: 'Is this one sort of wood or two?' As with veneered furniture, the timber exposed on the ends of a wooden object may be quite different from that seen on its face.

Seasoning timber

Wood is designed to carry sap, and so long as the tree that yields it is standing and alive its cells are full of water. From the moment that the tree is felled its timber starts to lose moisture, and the process of seasoning or drying begins. As the wood loses water, air moves in to fill the emptying spaces of its cells, and so it becomes lighter in weight. It also becomes harder and substantially stronger, and shrinks a little.

Seasoning continues slowly under natural conditions until a balance is reached between the water remaining in the wood, and the water vapour in the air around it. If you put a piece of wood in a hot oven, and leave it there long enough, you will get 'ovendry' timber which holds no water at all. But normally, even in a warm, centrally heated room, wood holds a quantity of water equal to at least 12 per cent of its own oven-dry weight. In a damper room, or out of doors, it holds considerably more.

This loss of water, with associated shrinkage and a slight change of shape, is something that man cannot completely check. All he can do is to recognize it, and take steps to lessen its ill effects. That is why wood is always seasoned or dried before use indoors. The easiest way to season wood is to stack it, out of doors, or in an open-sided shed, so that air can flow freely around every plank or piece. But this natural seasoning takes many months; it is usual for prime oak or ash plank to be left a whole year, or even longer, to season fully.

The alternative is kiln-seasoning, in which the planks are stacked on trolleys that are pushed into closed brickwork kilns, where the wood is slowly dried by artificial heat. This is a technical process requiring expert control to avoid too rapid drying and changes of shape. It proves very efficient and gives satisfactory seasoned timber in a matter of ten days or so.

Wood changes shape as it seasons because it shrinks by different amounts in different directions. Shrinkage in the longitudinal direction of growth is least, shrinkage along the radius of the log rather greater, and shrinkage round the circumference of the log greatest of all. The result of this difference in degree of shrinkage is best shown by two simple diagrams, Figs 2 and 3 below.

Fig. 2   The radial cut through a birch log, which splits it into two, becomes convex after seasoning

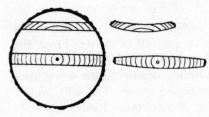

Fig. 3   *Top:* Boards cut from the side of the log become, after seasoning, concave on one side (originally the outer), and convex on the other (originally the inner). *Centre:* A board cut from the middle of a log acquires two convex faces during seasoning

One shows how the semicircular half of a log becomes convex across its 'straight' side after seasoning. The other reveals the corresponding changes in shape that occur in sawn planks.

Because of these alterations in shape and the general shrinkage of the material, it is essential to use thoroughly seasoned wood, finally seasoned *after* sawing, for all accurate work indoors.

Resistance to decay

One of the properties of wood that confuses many people, even after inquiry, is its resistance or liability to decay. It is natural to assume that if a wood is hard, strong, heavy or well seasoned, it will resist rot, and it is quite true that there are many hard, strong and heavy timbers that do resist decay indefinitely. But it is equally true that there are others equally hard, strong and heavy that decay rapidly when exposed to damp, even though they have been well seasoned beforehand.

Research has now made it clear that resistance to decay depends entirely on the chemical properties of a timber, not its physical ones. Rot is usually caused by fungi, occasionally by insects or marine borers. Certain woods – regardless of their strength, hardness or weight – hold chemical substances that are poisonous to invading organisms, other timbers do not. These natural chemicals are found only in the heartwood of the trees concerned.

The sapwood of all trees is technically 'perishable'. (See p. 35 for more about sapwood.)

Even perishable woods last indefinitely if kept very dry, for the fungi that cause decay all need some moisture for life and growth. Such woods may also last a long time at the bottom of a lake or the sea, because in that situation there is insufficient air for decay organisms to thrive. The test comes when wood is exposed to alternating dry and wet conditions – for example as a building timber or fence post.

Nowadays most timber that is to be used in situations where decay is likely, is treated first with a preservative chemical. This gives the most perishable woods, and even sapwood, a service life comparable to that of heartwood of oak – forty years or more. The effective chemicals usually stain the wood an unusual colour, either the dark brown of creosote oil or the yellowish green of the mixed salts of copper, chrome and arsenic that prove equally effective. If you have to identify a piece of wood used out of doors look out for these unusual colours, and make allowances for them.

Timber that is to be used indoors, especially furniture woods and fine veneers, is never likely to be exposed to conditions that favour decay. It will always be kept too dry. So you are unlikely to find any unusual colouring due to preservative treatment.

Surface finishes

Wood for decorative use, in furniture or as an ornament, is seldom left in its entirely natural state. It is usually treated with a wax, an oil, or possibly a modern plastic seal that will fill in its surface pores and give a smooth lustrous finish that can be readily polished. These substances usually contain some staining element that alters, to some degree, the natural colour of the wood. As a rule its grain is brought out more prominently, because the different tissues that make it up absorb the stain to various degrees. So it may prove easier to name treated timber than untreated material. Where doubts arise, it is usually a simple matter to find or expose an untreated surface. But make sure it belongs to the same piece of wood.

In this chapter we have travelled a long way from the first axeman felling a tree, and wondering about the curious pattern he found within, to decorative veneers and skilfully shaped timbers used in modern furniture. Yet the essential inner structure of the wood remains the same. Every timber preserves its own peculiarities, however it may be cut, shaped, elaborated or finished.

2 From Forest to Fireside

Great natural forests of birch and oak, pine and spruce, extend across North America, Europe and northern Asia. In the tropics one finds a tremendous variety of different but equally useful timber trees – teak, mahogany and rosewood – while Australia is the home of the magnificent eucalyptus. Despite all the clearings made for farming, or to harvest lumber, the bulk of the world's wood is still drawn from native forests rather than from man-made plantations.

Within the forest each tree springs up from naturally sown seed, in fierce competition with fellow seedlings arising about the same time. Only one wild tree in a thousand survives to maturity, the rest being crowded out or killed by disease, drought, insect attack, browsing deer or any of a score of hazards. Let us follow the course of life for a tree that *does* win through to full size and maturity. Year by year it grows steadily taller and stouter at the butt. The roots that draw water and mineral salts from the soil, and the leaves that win carbon compounds from the air around it, nourish it unfailingly. Each spring and summer it lays down a fresh annual ring of wood, outside its earlier rings. In years when it gets ample water and mineral elements from the soil, and its crown of leaves is active because it enjoys full sunlight, these rings will be wide. But in years of drought, or during periods when the crown is overshaded, they will be narrow. Each circumstance of the tree's career will be recorded in the wood of its stem.

Eventually, when it has grown tall and raised its shoots into the upper canopy of the forest, it flowers and bears seed. Once this process has begun, it is repeated every year, or every few years at least. Abundant seed, far more than is needed to replace natural losses, falls to the ground each year. Nature is always ready to renew the forest in this way, and when eventually the tree decays and dies through old age, there are always enough seedlings ready to grow taller to fill the gap.

By skilfully harvesting the maturing trees of the natural forests, before they age too far and start to decay, the forester can win a perpetual crop of timber from the woods without the need for replanting. The gaps will soon be made good through the process he calls natural regeneration. This method of 'selection felling' is widely practised in Switzerland and the neighbouring countries of the European Alps, because men dare not clear the mountainside forests for fear of avalanches.

In most of the world's forests, however, wide clearings are made when lumber is harvested, because modern machines can only operate efficiently and cheaply in this way. This means that foresters must renew the crop artificially, sometimes by sowing seed, but more usually by planting out young trees that they have raised in nurseries. Forest conservation has become the guiding principle in all the world's timber-producing countries. Those trees that are removed must be replaced with others to secure fresh harvests and profits for the future. By the same means the soil is saved from erosion by wind and weather, water supplies are maintained, wild life is safeguarded and the woods retain their unique scenic attraction for tourists.

The great American timber harvest

There have been times in the history of mankind when great forest clearances were made, and this course was sometimes the right one for the nation at the time. The early economy of New England rested heavily on the primeval woodland that awaited the axes of the settlers. By clearing the forests of oak, birch, maple and hickory, they won good land for plough and pasture – land that held the fertility arising from centuries under trees. At the same time they gained building timber, fencing timber, firewood and charcoal for iron-smelting. Ash from the burnt branchwood was rich in potash, a valuable fertilizer for the farm soil.

Lumber won from the primeval North American forests, which had cost the settlers nothing to grow, proved a valuable export. Britain, at that time, was starting to feel severe timber shortage, the outcome of her long neglect of forest conservation. While other European countries had kept their uplands under forest, the British had turned theirs into sheepwalks, believing that wool paid better than wood. As England's timber famine increased, her industrialists were ready to pay high prices for the prime American ash, white pine, hickory and oak that the colonists could easily fell in forests near the eastern seaboard. A steady trade grew up, bringing the New Englanders ample funds that could be used to buy manufactured goods from Europe. Britain is still the world's largest importer of lumber, and this traffic, though now directed largely to certain high-quality hardwoods that only the United States can produce, has gone on ever since.

At one point the British Government overreached itself. In legal theory the timber found on virgin forests in the new settlements belonged to the Crown, since the colonists had not planted it. During the eighteenth century the British Navy was desperate for masts for its oaken, sail-driven warships. Tall, straight American white pines, some growing nearly two hundred feet high, were ideal for this purpose. The British therefore sent surveyors to claim the best stems, and when a tree was chosen the surveyor marked it with a broad arrow – traditional symbol of Crown ownership. Naturally, the best trees were chosen, but the unfortunate settler on whose land they stood – and who could otherwise have sold them at a profit – was offered no compensation. This caused great ill-feeling and became one of the many grievances that led to the War of Independence.

When, in the nineteenth century, the Industrial Revolution began to change the American way of life, it created enormous demands on the country's forests, intensified by the parallel developments in Europe. By now the easily exploited woodlands of the north-eastern States had lost much of their best timber, so the big lumbering companies that sprang up concentrated their efforts on the next most readily available timber source – the Lake States. With steam sawmills now at their disposal, and railroads and steamers to carry their lumber towards eastern markets or overseas, the big concerns made short work of magnificent stands of pine, spruce and oak.

Once these northern woods had been overcut, the larger concerns that needed vast bulk supplies migrated to the south-eastern States. By the turn of the century, America's leading lumber-producing region was the land of the 'southern pines'. These were marketed collectively as 'pitch pine', and their hard, strong, resinous and brightly patterned timber was used throughout the growing cities and shipped on a great scale to Europe.

This resource, too, was eventually overcut, and the loggers moved over to the Pacific seaboard. Here they found great mountain wilderness woods of Sitka spruce and Douglas fir, western hemlock, western red cedar, and lodgepole pine, along with the grand and the noble firs, Port Orford cedar, ponderosa pine and the giant Californian redwoods. No greater natural resource has ever been presented to man, and it was only the distance of the West Coast from the eastern centres of consumption, before the opening of the Panama Canal, that had preserved the western forests from exploitation for so long. They are still the leading timber-producing region in America, and indeed in the whole world.

Early in the present century people began to realize that America's forest reserves were not, after all, inexhaustible. A pioneer conservationist, Gifford Pinchot, who had visited Europe to see how the Germans, French and Swiss managed to maintain perpetual supplies of timber despite centuries of felling, persuaded President Theodore Roosevelt to bring in forestry laws that now safeguard supplies for the future. Happily, most of the mighty western forests are still under Federal or State ownership. Felling can now be carried out only under regulations that require the land to be restocked.

As technology developed, a surprising range of methods was used to transport lumber from the stump to the sawmill. Horses and ox-teams were the rule during the early days in the east, supplemented by the floating of logs down rivers, or rafting across lakes. Then the railroads arrived, with quaint locomotives carrying huge spark-catchers above their chimneys, to lessen the risk of forest fire. Before the days of the motor truck, railroads played a large part in the logging of western forests too, but today huge articulated trucks, travelling over rough roads bulldozed and blasted from the mountainsides, have superseded them.

For the log's first journey from stump to roadside, horses, oxen and later tractors sufficed to bring the relatively small logs of the eastern forests over the fairly level ground that exists there. But in the western mountains the land is too rough and steep, and the trees too large, for surface transport to work well. So the loggers developed cableways and skidding systems, which progressed from simple wires slung between a winch and a distant tree, to elaborate 'high-lead' arrangements that are a major feat of engineering. Their efficiency formerly hung on the use of a spar tree, a tall, selected specimen that carried the main cables. This tree was climbed by a daring rigger who topped it and then fixed guy-ropes and pulleys. Today the main support for the cables is usually a portable steel tower, or even a captive balloon that gives the necessary 'lift' to the haulage system.

The harvesting of America's vast virgin forests was the most spectacular and rapid move of its kind that the world has seen. It gave the developing country great capital resource and prompted rapid advances in transport engineering and timber technology. It left its problems, and it is odd to reflect that much of the land first cleared by New England settlers is once again under trees – for it proved less profitable to farm than the prairies.

Similar developments have taken place in all those countries that possess temperate-zone forests of pine, spruce and larch among the softwoods, or the leading temperate hardwood trees – oaks, maples, elms, beeches, birches and ash. Everywhere harvesting has been speeded and cheapened by modern machines, and thought is now given to restoring a fresh tree crop. Scandinavia, Russia and the Central European countries show parallel patterns, and Britain's timber shortage is being remedied to some degree by a big afforestation programme.

Harvesting tropical timbers

Tropical forests show a very different picture from that shown by those of the temperate north, where there are relatively few trees and all are marketable for some purpose or another. The hot, moist jungles of Central and South America, Africa, India and South-east Asia hold a bewildering range of tree species – some useful to man, others not. The needs of the inhabitants do not require them to exploit their forests much for their own use. Poles for building, and palm leaves that provide wall-cladding and roofing material for their simple dwellings, are easily cut from the lower trees or bushes. A great jungle monarch is used only for some special purpose, such as the construction of a big dug-out canoe. Yet clearings are often made for cultivation, because many tropical peoples practise 'shifting cultivation'. They clear a patch of jungle, burn its timber and branchwood, grow crops for a few years on land enriched with wood ash, and then move on. The jungle trees quickly return, by natural seeding, to reclaim the cleared land as their own.

When the first explorers and traders reached Africa, India and South America, they found that the native peoples were skilled in wood-cutting but had limited uses for the wonderful timbers that grew to great sizes in their jungles. Indian teak was an exception, for it had long been used in shipbuilding, house-building and as furniture.

Here lay a resource that could quickly be made to yield a profit. The trees had grown at no cost, and could be bought for a trifling royalty from the headmen, chiefs or rajahs who were their nominal owners. Africans, Asiatics and South Americans were ready to work as tree-fellers and hauliers for low wages, as a cash supplement to their subsistence farming. Timber provided a sound return cargo for ships that had brought trade goods from Europe or North America. Ready markets existed for choice woods having unusual colours or attractive figure.

On this sound economic basis world trade in tropical timbers began, and it has prospered ever since despite changes in fashion, politics and the economic conditions in the countries concerned. It has proved of great value to the exporting countries, needing foreign exchange. Where progressive governments exist, the yield from the forest is regulated by law so that it can be maintained far into the future. Virtually all commercial tropical timbers are hardwoods, and most can be described, in the broad sense, as luxury woods. They are used for good furniture, high-grade joinery, veneers and decorative work, seldom for everyday construction. Because they have to be sought in distant, roadless forests, and carried by sea over thousands of miles, they are never cheap.

Merchants who deal in tropical timbers have first to seek out mature trees of desirable kinds in remote jungles. Only one tree in a hundred is likely to be of the right sort and size. Each is recognized by key features of bark, leaf, flower or fruit, and general appearance. Selected boles are given a distinctive mark, usually a blow with a specially patterned hammer. Certain timbers, including teak, are too heavy to float when freshly felled. These are killed by ring-barking, and left for a whole year, during which time their doomed crowns of foliage transpire the sap from the mighty trunks; this makes them light enough for floating.

The traditional method of felling is with hand tools, but even in the tropics the modern power-saw is gaining ground. Transport, too, was originally done by the muscles of man or beast, or by floating down waterways, but powerful crawler tractors are nowadays busy in many tropical jungles. Certain trees form enormous buttresses at their base, and even with modern power-tools it is barely possible to fell them at ground-level. Springboards are therefore driven into the tree, perhaps twelve feet up from the ground, and the fellers stand on these to cut through the great trunk at a still higher level.

Often sheer manpower, cheap if wasteful, is used to haul the logs with ropes over rough rollers towards the nearest waterway. Buffalo-teams and ox-teams are employed in some countries, while in Burma and parts of India trained elephants are used both to haul and to stack timbers. Few tropical lands have adequate roads or railways, so transport to the seaports is usually done by river. Some tropical woods are so heavy, even after seasoning, that they cannot float. These are supported on rafts of lighter timbers or bamboos, or else loaded into river barges.

Some tropical hardwood is shipped in the form of round logs. This is the usual practice for veneer timbers, which will be 'peeled' on some great lathe after they have reached their destination. But much is sawn into square baulks to save shipping space. Once this was done by hand, by means of huge pit-saws. Today modern sawmills equipped with bandsaws operate at the main tropical ports of shipment.

The woods that North American and European importers seek from the tropics are naturally those that they cannot match from home sources, either for quality or for cost. But a good deal depends on fashion, and on techniques for handling unusual materials. During the eighteenth and early nineteenth centuries there was a tremendous vogue for mahogany. Its immediate appeal lay in its glowing red colour, but it also attracted the craftsman cabinet-maker because it is soft enough to be easily carved, hard enough for everyday wear, light in weight and remarkably stable. Once you have worked a piece of seasoned mahogany it will never shrink or warp. It was favoured by the wealthier classes because nothing like it grows in the cold north. A yeoman farmer might – in fact often did – boast a handsome oak chest or chair, but only the rich could afford the fashionable elegance of imported mahogany. Then there were fascinating variations between one piece and the next. Wood cut from crutches or branch intersections showed beautiful patterns; at an early date this was discovered and such logs were sawn by hand into rather thick but very attractive veneers.

True mahogany comes only from the Caribbean region, notably from Honduras and Guatemala. Accessible forests were soon overexploited and merchants looked elsewhere for supplies. Several African trees were found to yield timbers that would pass in trade as 'mahoganies', though some were not related to the Central American kind and had different microscopic structures and working properties. The marketing of one timber by the name of another is common practice in the tropical hardwood trade. It is deplored by the botanists, and it confuses the student, yet it is hard to deny its commercial common sense.

Honduras mahogany, African mahogany and sapele – a timber also sold as a 'mahogany' – are featured in this book, and the reader will be able to compare and contrast their characters.

Today mahogany, though still extensively imported and applied, is often considered a timber for the collector of antique – or at least of period – furniture. We acknowledge, regretfully, that the best was made a hundred years ago. The current fashion is for brightly coloured woods that can be effectively applied, as thin veneers, to radios, television sets, car fascias, cocktail cabinets and similar pieces of specialized 'furniture' that – people feel – need something gay yet natural to offset their strictly technical purpose and design. The contrasting tiger stripes of zebrawood, the luscious mauve of purpleheart, and the intricate light and dark veining of sweet-scented rosewood, all accord with this mood.

This chapter has ranged far and wide, from the seedling tree in a northern forest to the sturdy loggers of the Pacific slope and the exotic beauty of tropical woods wrought by craftsmen into artistic furniture. But the study of timber runs that way. Every piece you possess has flourished in the green heart of a forest, and been prepared for your use by sturdy loggers and able technicians. Tracing it back to its source will open wide horizons, and your search must begin with the critical question: 'What wood is that?'

PART II: WOOD IDENTIFICATION

3 Wood Formation and Structure: The Clues to Identity

To tell timbers apart by using the simple keys that follow you must first look closely into the structure of wood in general. You will then be able to pick out the essential tissues, such as rings, rays and pores, more readily, because you will know the purposes they serve and the way they were formed within the living tree.

Trees form wood in their tall, slender stems to support their crowns of foliage and to supply their leaves with sap. It is this dual purpose that makes wood such a fascinating material to use, study and admire. Every piece of wood is both firm and fibrous, solid to the touch yet porous to air and liquids, strong but easily shaped. No man-made material can match its intricate patterns of grain and colour, or prove so adaptable to every kind of structural and decorative use.

Cambium, bast and bark

The youngest shoots of a tree are green and soft like those of a smaller plant. They carry sap upwards from the roots, and also downwards from the leaves, through clusters of fine conductive cells called *vascular bundles* (Fig. 4). But this arrangement lasts for only one year. As soon as growth begins in the second spring, the scattered bundles round the edge of each one-year-old shoot become united into a cambial ring or *cambium* (Fig. 5). This cambium forms a sheath of very thin but very active cells. It lies just below the bark of every living tree-stem, from ground-level right up to the tops of the highest branches. It is so thin that it can only be seen through a microscope, yet it produces, on its inward side, the whole woody substance of the tree's trunk and lesser stems.

Outside the cambium there are two very thin, though vital, tissues round every tree-trunk. One is the *bast*, which carries a small downward flow of sap, rich in sugar, from the tree's leaves to feed the cambium and the roots. The other is the *bark*, a protective layer that is created and renewed by its own thin *bark cambium*.

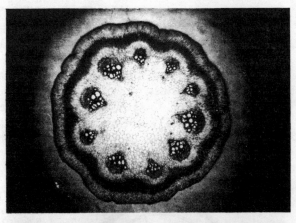

Fig. 4 Cross-section of a one-year-old twig of mock-orange or syringa (*Philadelphus coronarius*), with pear-shaped, radially disposed vascular bundles. Running through the centre of each of these and joining them all up are the tiny cambium cells, which show as a very faint line

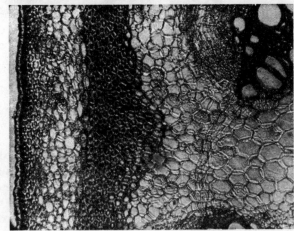

Fig. 5 Part of the same cross-section magnified further to show the thin line of cambium cells. This runs down obliquely from the centre of a vascular bundle (*top right*) to the next bundle below. Bast and bark lie to the left, wood will form on the right of the cambium

The bark shields the bast and the delicate cambium from many kinds of possible harm. It resists attacks by gnawing squirrels or browsing deer, lessens damage by chance knocks, sudden sharp frosts and scorching sunshine, and helps to keep out harmful fungi and insects. Its corky layers keep the growing wood within continually moist. In a living tree there is scarcely any loss of water through the bark, and even the air the stem needs to breathe must enter through special pores called *lenticels*. As the tree ages, the bark gets steadily thicker. Its pattern then varies from one kind of tree to another, and this can prove a great help in identification (Key 10).

Wood can only be formed in a living stem by a cambium that is nourished by an outer layer of bast and protected by a second, outermost layer of bark (Fig. 6). All the time it is growing there must be leaves in the tree's crown busy winning carbon dioxide from the air, to be transformed by the magic chemistry of growth into the new carbon compounds that compose the wood. And all the time, too, sap must be actively flowing up the previously formed tissues of the stem, from the roots busy exploring the soil below, to carry mineral salts needed for the life processes of the tree. Therefore, wood is formed, in temperate climates, only during the spring and summer months, during the active seasons of the tree's life.

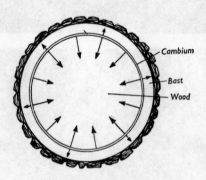

Fig. 6 Cross-section of a twig with bark, bast, cambium and wood, showing how the cambium builds up bast towards the *outside* and wood towards the *inside*

Cambium

Bast

Wood

We shall see later how the wood formed in summer differs from that formed in spring. No wood at all is formed during autumn, when the rhythm of growth slackens, nor in winter, when the tree rests. Many trees, of course, are leafless at that time of year.

Both bark and bast are stripped away when timber is felled, or soon after, and the student or user of wood is concerned only with the *true wood* laid down *within* the cambium ring. This true wood consists of two kinds of tissue, the *rings* and the *rays* (Keys 3 and 6). As the names suggest, the rings form circles round the centre of the tree-trunk, and because one ring is normally formed each year they are called *annual rings* or *growth rings* (Figs 7, 16, 17, 18).

The rays, which radiate outwards from the centre of the trunk (or follow the same pattern even if they do not reach the centre) make up a much smaller proportion of the tree's substance. But they are very useful aids to identification, and can also be highly decorative. Broadly speaking, the rays serve as storage tissues, and also carry food substances inwards or outwards through the living trunk. The rings, on the other hand, give strength and transport sap *upwards*.

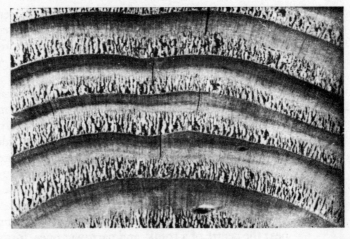

Fig. 7 Cross-section of Douglas fir, showing obvious annual rings, with pale, light springwood and dense, dark summerwood

Springwood and summerwood

All who know the woods of pine, spruce, birch or oak in America and other northern countries will realize that each spring there is a great outburst of active life. New shoots appear, clad in fresh green leaves, and these make a heavy demand on each tree's roots and stem for greater supplies of sap. The cambium of the tree responds by laying down a sheath of new wood, made up of elements that have thin walls and a large central space—in brief, efficient tubes. The circular band of wood that results is called *springwood*. It is light in colour, light in weight when seasoned, and only moderately strong.

When summer succeeds spring, the cambium suddenly changes the character of the wood it produces. Most of the sap the tree needs can now flow upwards through the springwood, but the stem needs more structural strength because it has grown stouter and heavier. So, during the summer months, the cambium produces fibres with thick, strong walls, and small cavities. The resulting *summerwood* is darker in colour, heavy even when seasoned, and very strong.

Because this pair of layers of different types of wood is formed on a circular plan every year, it makes up one *annual ring*. In many timbers these rings are easily seen with the naked eye. Count them on the cross-section of any tree or log, and you know its age. In the same way, if you count the rings on the stump of a felled tree, you can say exactly how long that tree grew before it was cut down. The oldest and smallest ring is always at the centre, the youngest and largest ring at the outside (Fig. 7).

The rings appear as true circles only when the stem is cut across. If it is cut in any other way they will show as curves, straight bands, or irregular patches of tissues that contrast in colour, density, strength and hardness. This variation gives the main *grain* or *figure*, and aids identification (Key 5).

Though rings are always formed in the same *way*, they are not always the same *width*. Actively growing trees form wide rings, slowly growing trees form narrow ones. In a forest, you may find two trees of the same kind and age, growing side by side. But one is well placed to get sunlight and is therefore growing fast and forming broad rings, whereas the other is overshaded and growing slowly, forming narrow rings. Again, the same tree may make broad rings during its vigorous youth, but narrow ones in its old age. It is quite common, indeed usual, to find broad rings near the centre of a trunk, and narrower ones, formed later in life, farther out.

It follows from this that the *width* of the annual ring is of no use at all for the work of identifying timbers. But it is of great importance to the users, who will, for example, select *close-ringed* or *fine-grained* timber for fine joinery, yet accept *wide-ringed* or *coarse-grained* timber for structural jobs. Timber-merchants grade their stock to match these needs. For instance, they import fine-grained spruce from northern Canada or Scandinavia for ladder-makers, but direct coarse-grained spruce from forests farther south to customers who manufacture packing cases or make flooring.

Knots

As a tree steadily expands in girth, year by year, it slowly includes the base of each branch within its growing trunk. When eventually the trunk is sawn up, buried branches are usually exposed to view, and we call them *knots*. In a cross-section of a stem, they run out radially; but if the knots themselves are cut straight across, they show up as dark circles, while a radial cut reveals them as curious oval shapes, ending abruptly at the point where the branch fell off or was pruned away (Fig. 8).

Fig. 8 Two knots seen on the cross-section of a tree's stem; knots are buried branches

All trees form knots near their centres, but as the years go on and the smaller side branches fall away, a tree may eventually lay down layers of wood that lie farther out than its longest remaining lower branch. This wood is called 'knot-free' or 'clear'. It is easier to work than wood that holds knots, and has more consistent strength properties, so it is more valuable.

Knots seldom help you to identify timbers for there are knots hidden somewhere in every tree. It is a matter of chance whether you get a piece of wood from its knotty heart or its knot-free outer zone. Many wooden objects are so small that they can be cut from the 'clear' stretch of timber that lies between two knot clusters. For these reasons knots should normally be disregarded in your search for the name of a tree. Occasionally, as with Bird's-eye maple, they provide a valuable clue. Pine trees can be distinguished from other conifers by the arrangement of their knots, which are *all* in clusters. In the other conifers some knots are clustered but others arise separately.

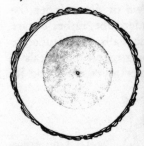

Fig. 9 Cross-section of the stem of a tree with well-defined heartwood. The stem has dark heartwood surrounded by pale sapwood and an outermost layer of bark

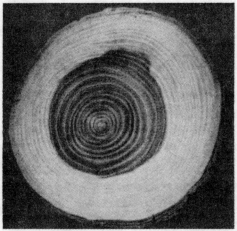

Fig. 10 Cross-section of an elm log, a typical example of dark heartwood and pale sapwood. Note how the boundary follows the trend of the annual rings, but does not always coincide with them

Heartwood and sapwood

All wood is formed to carry sap. Therefore, it all starts life as *sapwood*. As a trunk or branch gets older and stouter, the inner core is no longer needed for this purpose, though it is still required for structural support. Slow chemical changes then take place in the inner layers, which become *heartwood*.

35

Heartwood is essentially wood that has ceased to carry sap. It is usually harder than the sapwood that surrounds it, and somewhat stronger. In coniferous trees, or softwoods, it is often resinous, while in the broad-leaved trees, or hardwoods, it may hold abundant tannins. Once harvested, it is often more resistant to decay than the sapwood, because these chemicals prevent the growth of fungi. But this resistance is found only in certain timbers such as oak and western red cedar, and is not universal.

Though heartwood can be found in the stouter stems of every sort of tree, it is not always obvious by its colour. If a clear colour difference is apparent, this is a clue to identity, as shown in Key 13, Sapwood Definition, and Figs 9 and 10.

Many trees show a dramatic colour difference between sapwood and heartwood which makes recognition easy. The common laburnum, for instance, has a dark, lustrous, blackish-brown heartwood, surrounded by pale yellow sapwood. In contrast, ash darkens only slightly, and only a close examination will show where its heartwood begins.

SOFTWOOD TIMBERS

The wood of the coniferous or needle-leaved trees, which is called *softwood* because it is, as a rule, relatively soft and easily worked, has a simple structure (Fig. 11). Each ring consists mainly of a single type of cell, the *tracheid* or *fibre*. Under the microscope this is seen to be a long, slender tube, rather square in cross-section, with a hollow cavity. The walls of the tube are made up of two carbon compounds or organic chemicals – flexible *cellulose* and rigid *lignin*. The former gives the wood its

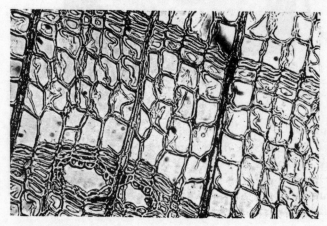

Fig. 11 Cross-section of larch, a typical softwood, seen under a high-powered microscope. Large spring-wood tracheids are followed by small summerwood tracheids, forming together an annual ring. Narrow rays run across this ring, and two resin canals are seen, lower left

toughness, while the latter contributes firmness and solidity. The centre of the tube holds a small quantity of living tissue or *protoplasm*, but is otherwise nothing more than space for the transport of sap. As this sap ascends the tree-trunk, it is passed from one fibre to another through minute openings known as *pits*.

Rays and resin in the softwoods

There are two other features that help to mark out most, though not all, softwoods. The rays that run at right angles to the annual rings, and therefore radiate out from the centre, are narrow and scarcely visible even under a lens. They never show as an attractive figure on exposed surfaces, and their small size leads one to suspect that an unknown timber *may* be a softwood.

Resin, on the other hand, is found in softwoods but never in other timbers. It is formed in tiny canals that run through the wood in several directions. They are too small to see without a microscope, but most softwoods can be named as such by their characteristic 'turpentine' smell. This is most obvious on freshly cut surfaces. It may disappear as time goes on.

HARDWOOD TIMBERS

The wood of all those trees that bear broad leaves, rather than narrow needles, is called *hardwood*. In most species it is physically harder than the softwood formed by conifers, and also denser; but there are exceptions such as the soft, light woods of the poplars. The general pattern of hardwood timbers resembles that of softwoods, but important differences make it distinct and more complex. Luckily this complexity aids identification (Fig. 12).

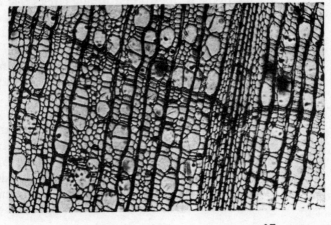

Fig. 12 Cross-section of alder, a typical hardwood, seen under the microscope. The larger openings are pores or vessels, the smaller ones tracheids. Alder is diffuse-porous. Summerwood is not well defined, but the end of an annual ring can be seen. The dark lines that cross this are rays

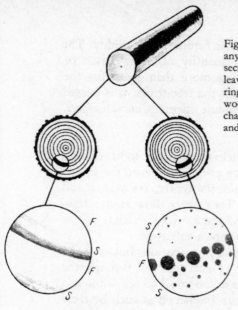

Fig. 13. *Left:* Cross-section of a wood without any pores; a conifer or softwood. *Right:* Cross-section of a wood that holds pores; a broad-leaved tree or hardwood – in this case a ring-porous one. F = Springwood. S = Summerwood. This seasonal variation in wood character occurs in all temperate-zone timbers and results in their annual rings

Fig. 14 Vessels, or pores, show as tiny furrows or striations on the surface of many, though not all, hardwoods. They never occur in softwoods

Vessels in the hardwood annual ring

When a broad-leaved tree expands its fresh foliage in spring, it needs an even faster upward flow of sap than does a conifer. Besides forming tracheids or thin, tube-shaped cells like those of a conifer, it builds up special conducting channels called *vessels* or *pores*. Each vessel is composed of many cells, and most vessels are so large that they are easily seen with the naked eye, or very easily with a hand lens. On a cross-cut surface they show as tiny pin-holes (Fig. 13). On other surfaces they can be seen as fine lines or striations (Fig. 14 and Key 4). In certain timbers the vessels arise mainly in the springwood, forming circles that coincide with the annual rings. These hardwoods are called *ring-porous*.

Other trees have more numerous, smaller vessels scattered through both springwood and summerwood. These are called *diffuse-porous* woods. Each pattern is constant for any particular kind of tree, and can prove a great help in identification (Fig. 15).

Obvious and obscure hardwood rings

Most hardwoods are formed by broad-leaved trees that grow in regions where there is a marked difference between winter and summer, or between a wet season when growth is possible and a dry season when it ceases. Trees that grow under such climates, even in the tropics, have a regular annual rhythm of growth, and produce wood with obvious annual rings (Figs 7, 16).

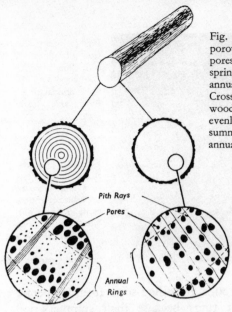

Fig. 15 *Left:* Cross-section of a ring-porous hardwood, that is, one with large pores arranged in obvious bands in the springwood, so that the boundaries of the annual rings are clearly visible. *Right:* Cross-section of a diffuse-porous hardwood, that is, one with large pores spread evenly through both springwood and summerwood; here the boundaries of the annual rings can only just be picked out

Pith Rays
Pores
Annual Rings

Fig. 16 Cross-section of elm, a semi-diffuse-porous hardwood, showing obvious annual rings. The springwood of each ring is narrow and pale, with obvious large pores, while the summerwood is broader and darker, with smaller pores. Heartwood is seen at lower left

But in some tropical countries near the Equator the weather remains hot the whole year round. The rains fall at all times of year, too, and growth never stops. Under these conditions the trees, which are naturally evergreen, do not form distinct annual rings, though the microscope usually reveals a yearly pause. A number of these 'ringless' timbers are sold in the commercial market.

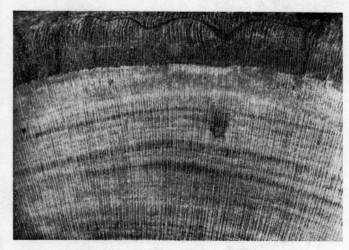

Fig. 17 Cross-section of plum. A fully diffuse-porous hardwood with obscure annual rings. Rays are clearly seen

Fig. 18 *Opposite:* A cut across a log or a long piece of sawn timber exposes its cross-section. Note annual rings – sometimes seen complete, sometimes only partially ▶

Another group of hardwoods, including some from temperate lands, do not show clear rings because their summerwood resembles their springwood (Fig. 17). Key 3, Rings, makes use of all these features.

Ray figure in hardwoods

Many hardwoods develop large and characteristic rays, which are exposed by skilful cutting to show on the finished surfaces of woodwork as plates or bands. The silver grain of oak is a familiar example. In other hardwoods the rays, though individually small, have a colour that contrasts with the rest of the timber. Beech, for example, can easily be picked out by the dark brown flecks – each fleck a separate small ray – that show up against its pale brown background.

Other hardwoods again have thin featureless rays that cannot be seen by the naked eye. This characteristic, though negative, also helps people to name the woods concerned. Key 6, Rays, makes use of these characteristics.

Other features of hardwoods

Hardwoods do not hold resins, so the absence of any 'turpentine' smell can prove a useful pointer.

But many hardwoods hold chemicals that give off, from freshly felled or newly worked timbers, very characteristic odours. The sharp 'tannic-acid' tang of oak is familiar and unmistakable, and so are the 'mouldy' or 'earthy' smell of elm (Key 9) and the delightful 'blossom' smell of rosewood.

How structure is revealed on wood surfaces

Owing to the way in which wood is built up, it never appears the same on different surfaces. A cut *across* the tree-trunk, or 'across the grain' of any piece of sawn timber, will expose its annual rings clearly as circles, but the rays can only show as lines or narrow bands (Figs 18, 20).

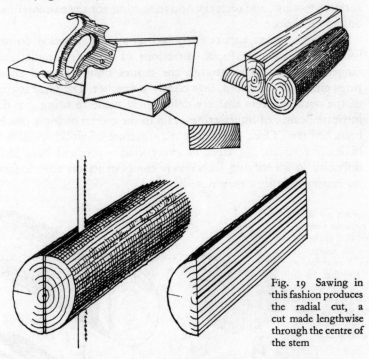

Fig. 19 Sawing in this fashion produces the radial cut, a cut made lengthwise through the centre of the stem

Conversely, a cut down the length of a log, which passes through its centre, will expose the full surface of any major *rays*, so that they show as flat plates. But on this *radial cut* the annual rings will show as straight lines, or bands, not circles (Fig. 19).

Woods with prominent *ray figure* are often cut in this way to show it off to best advantage. Decorative veneers of oak, for example, are usually so cut to reveal its 'silver grain' (Fig. 20).

If a longitudinal cut is made farther out, away from the centre of a log, another pattern will appear, for the annual rings will now show as bold stripes, or patches, while the rays will be narrow. In certain timbers, such as Douglas fir, bold and attractive patterns are revealed as 'slash grain' by this *tangential cut* (Fig. 21).

Allied to slash grain is the figure on a *rotary-cut* veneer (see p. 15), made by revolving a large log in a lathe and peeling off thin outer layers as a continuous strip (Fig. 22).

Veneers – the thin decorative layers that are glued to furniture as an ornamental finish – can show only a limited range of wood characters, though this is usually ample for identification. But any larger solid piece of wood will present two or three different surfaces to view, and each can add fresh hints for the naming of an unknown timber.

In practice many surfaces exposed on a piece of wood do not follow the classic textbook directions of cross-cut, radial and tangential. They pass through the tissues obliquely, and show rings clearly at one point, rays clearly at another. The same is true of the *crutch* veneers that are deliberately made to bring out the intricate beauty of intersecting grain in the crown of some much-branched tree. Once you know the structure of wood – which is basically constant for trees of every kind – you will have little difficulty in identifying its various components on any surface, no matter how they may run.

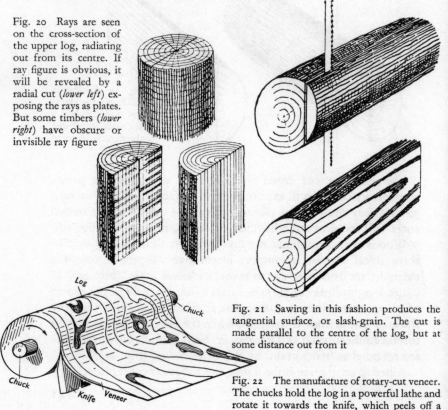

Fig. 20 Rays are seen on the cross-section of the upper log, radiating out from its centre. If ray figure is obvious, it will be revealed by a radial cut (*lower left*) exposing the rays as plates. But some timbers (*lower right*) have obscure or invisible ray figure

Fig. 21 Sawing in this fashion produces the tangential surface, or slash-grain. The cut is made parallel to the centre of the log, but at some distance out from it

Fig. 22 The manufacture of rotary-cut veneer. The chucks hold the log in a powerful lathe and rotate it towards the knife, which peels off a sheet of veneer

4 Keys for Naming Timbers

In practice, people identify timbers by looking at several characters of each specimen. One single feature is seldom enough. Although one striking character of colour, grain, weight or hardness may at once suggest a name, it is always advisable to find other confirmatory characteristics.

The keys that follow are designed to enable you to seize upon any striking characteristic of any piece of wood, and follow up its naming from that starting-point.

The first key is based on General Colour, which is the best feature to start with for most woods. All the succeeding keys are cross-referenced to this one. The full range of the keys is as follows:

Key 1 General Colour. The main key (p. 44).
Key 2 Secondary Colour – of heartwood (p. 59).
Key 3 Rings, i.e. annual rings or growth rings (p. 62).
Key 4 Pores (p. 63).
Key 5 Grain on Longitudinal Surfaces (p. 65).
Key 6 Rays (p. 66).
Key 7 Hardness (p. 67).
Key 8 Weight, i.e. specific gravity or density (p. 68).
Key 9 Smell (p. 70).
Key 10 Bark (p. 71).
Key 11 Leaf Shape (p. 72).
Key 12 Country of Origin (p. 73).
Key 13 Sapwood Definition (p. 75).
Key 14 Class of Use (p. 75).

Keys 1–8 apply to all the forty timbers in this collection. Keys 9, 10 and 11 include only those woods that are likely to present useful features of smell, bark pattern or leaf shape. Key 12 applies only when the country of origin of a sample of timber is known. Key 13 sets out the possibilities where a specimen shows clearly defined sapwood. Key 14 ranks all the woods into broad classes of use.

It is emphasized that you can start your identification with *any* of the keys that appear to apply. If a wood strikes you as very heavy, for example, you can take Key 8, Weight, first. If you have a leaf from the tree, then Key 11, Leaf Shape, is a good starting-point, and so on.

GENERAL COLOUR

This is the main key, and all other keys are linked to it. As a glance at the specimens will show, most woods are quite easy to divide into groups on the basis of their general colour. Though wood colours are seldom vivid, only a minority of timbers can truly be described as brown. Many are paler to varying degrees, and can easily be grouped as yellowish or whitish. Others contain characteristic pigments that make them reddish, purplish, greyish or even black.

Allowance must be made for the fact that most timbers darken slowly through exposure to sunlight and air. Surface treatments, including the so-called 'clear' varnishes, also tend to render wood darker. Both these changes affect only the surface layers, and it is usually easy to find or reveal an unaltered surface that has neither been treated nor been rendered darker through long exposure.

There are six possibilities in Key 1, as follows: A. Wood is whitish (pp. 44–46); B. Wood is yellowish (pp. 46–49); C. Wood is purplish or crimson (pp. 50–51); D. Wood is reddish or pinkish (pp. 51–54); E. Wood is brownish (pp. 54–58); and F. Wood is blackish or greyish (p. 59).

A: Wood is whitish

A I ASH

Group features:	One-coloured
	Annual rings distinct
	Pores in rings, coarse
	Grain distinct
	Rays obscure
	Hard
	Heavy
Peculiarities:	The strong pattern of large pores following annual rings is unmistakable. Sapwood not defined.

A 2 ASPEN POPLAR

Group features:	One-coloured
	Rings obscure

A 2 *(cont.)*   Pores diffuse, fine
Grain obscure or invisible
Rays obscure
Soft
Very light

Peculiarities: Aspen poplar is exceptionally soft and light, spongy, fibrous and rough-surfaced, but otherwise featureless. Whitish or very pale yellow. Sapwood not defined.

A3 AVODIRÉ

Group
features: One-coloured
Rings obscure
Pores diffuse, fine
Grain faint
Rays obscure
Hard
Light

Peculiarities: Avodiré shows faint but definite grain and is tougher, heavier and distinctly harder than aspen poplar. White to pale yellow or biscuit colour. Sapwood not defined.

A4 BIRD'S-EYE MAPLE

Group
features: Two-coloured
Rings distinct
Pores diffuse, fine
Grain distinct
Rays distinct
Very hard
Heavy

Peculiarities: The peculiar 'bird's-eye' figure, with the grain forming circles round small dark knots, distinguishes this attractive white timber. Sapwood ill-defined.

Group
features:

One-coloured
Rings distinct
Pores diffuse, fine
Grain distinct
Rays distinct
Very hard
Heavy

Peculiarities: Yellowish white, with darker summerwood, harder and heavier than sycamore maple, lacks the circular figure of Bird's-eye maple.

A 6 SYCAMORE MAPLE

Group
features:

One-coloured
Rings obscure
Pores diffuse, fine
Grain faint
Rays distinct
Hard
Light

Peculiarities: Rings and grain less distinct than in maple and somewhat lighter, pale greyish white. Sapwood ill-defined.

B: Wood is yellowish

B I ANTIARIS

Group
features:

One-coloured
Rings obscure
Pores diffuse, coarse
Grain obscure or invisible
Rays obscure
Soft
Very light
Smell: unpleasant

B1 *(cont.)*

Peculiarities: A featureless, dull-surfaced wood, with large pore lines evident on longitudinal surfaces. Sapwood not defined.

B2 BIRCH

Group
features: One-coloured
Rings obscure
Pores diffuse, fine
Grain faint
Rays obscure
Hard
Heavy

Peculiarities: Birch always shows dull surfaces, lacking shine, lustre or other conspicuous features. Pale brownish yellow. Sapwood not defined.

B3 LACEWOOD PLANE

Group
features: One-coloured: pale brown
Rings distinct
Pores diffuse, fine
Grain faint
Rays distinct
Hard
Light

Peculiarities: The lively dappled effect of the rays, suggested by the name 'lacewood', is unmistakable. Sapwood not defined.

B4 LIME

Group
features: One-coloured
Rings obscure
Pores diffuse, fine
Grain obscure or invisible
Rays obscure
Soft
Light
Smell: pleasant

B4 *(cont.)*

Peculiarities: Softness, combined with a mild lustre and a pleasing smell when freshly cut, mark out lime from similar yellowish timbers. Sapwood not defined.

B5 OAK

Group features: One-coloured (two-coloured on radial surfaces)
Rings distinct
Pores in distinct rings, coarse
Grain distinct
Rays distinct
Hard
Heavy
Smell: strong, tannic acid

Peculiarities: Strong grain on all surfaces. Distinct contrasts between pale rays and darker ring tissues make radial surfaces appear two-coloured. Sapwood pale yellow, sharply defined.

B6 PINE, PONDEROSA

Group features: Two-coloured
Rings distinct
Pores absent
Grain distinct
Rays obscure
Soft
Light
Smell: resinous, resin canals present

Peculiarities: Colour contrast between red-brown summerwood and yellow springwood of each ring. Sapwood creamy white, clearly defined.

B7 PRIMAVERA

Group features:	One-coloured
	Rings distinct
	Pores diffuse, coarse
	Grain distinct
	Rays distinct
	Hard
	Very light

Peculiarities: Lustrous surface, recalling that of mahogany, distinguishes primavera from similar yellowish woods. Sapwood not defined.

B8 SATINWOOD, EAST INDIAN

Group features:	One-coloured
	Rings obscure
	Pores diffuse, very fine
	Grain obscure or invisible
	Rays obscure
	Hard
	Very heavy

Peculiarities: The very lustrous surface that gives this clear yellow wood is name is distinctive. Sapwood not defined.

B9 ZEBRAWOOD

Group features:	Two-coloured
	Rings distinct
	Pores diffuse, coarse
	Grain distinct
	Rays obscure
	Hard
	Heavy
	Smell: unpleasant

Peculiarities: Striking contrast between yellow and brown bands following trend of annual rings. Sapwood pale yellow.

C: Wood is purplish or crimson

C1 PADOUK, ANDAMAN

Group features:	Two-coloured
	Rings distinct
	Pores in rings, coarse
	Grain distinct
	Rays obscure
	Hard
	Very heavy
	Smell: pleasant

Peculiarities: Rich crimson to purplish colour with darker or blackish streaks; pores in rings. Sapwood whitish to yellowish grey.

C2 PURPLEHEART

Group features:	One-coloured
	Rings obscure
	Pores diffuse, fine
	Grain faint
	Rays obscure
	Hard
	Very heavy

Peculiarities: Uniform purple or violet colour, density, lack of visible texture. Sapwood whitish.

C3 BUBINGA

Group features:	Two-coloured
	Rings distinct
	Pores diffuse, fine
	Grain faint
	Rays distinct
	Hard
	Very heavy

C 3 (cont.)

Peculiarities: Purplish-brown or crimson-brown ground colour with deeper bands or mottling. Sapwood white.

Note.   Some specimens of Black American walnut (E 11) are so purplish in colour that they may be considered here. This timber can be distinguished from purpleheart and Andaman padouk by its distinct rings, and from bubinga by its even, chocolate to purplish-brown colour.

D : Wood is reddish or pinkish

D I AGBA

Group features:
One-coloured
Rings distinct
Pores diffuse, coarse
Grain faint
Rays obscure
Soft
Light
Smell: peppery

Peculiarities: Brick-red to terracotta-pink in colour. Sapwood not defined.

D2 BEECH

Group features:
Two-coloured
Rings obscure
Pores diffuse, fine
Grain faint
Rays distinct
Hard
Heavy

D 2 *(cont.)*

Peculiarities: Chocolate brown rays show everywhere as small flecks or plates on a pinkish-brown ground. Sapwood not defined.

D 3 CEDAR, WESTERN RED

Group features: Two coloured
Rings distinct
Pores absent
Grain faint
Rays obscure
Soft
Very light
Smell: aromatic

Peculiarities: Absence of vessels. Even red-brown grain and colours, with only low contrast between springwood and summerwood. Odour. Sapwood creamy yellow.

D 4 DOUGLAS FIR

Group features: Two-coloured
Rings distinct
Pores absent
Grain distinct
Rays obscure
Soft
Light
Smell: resinous, resin canals present

Peculiarities: Strong contrast between hard red-brown summerwood and soft yellowish-pink springwood. Sapwood yellowish white.

D 5 MAHOGANY, AFRICAN

Group features: One-coloured
Rings obscure
Pores diffuse, grouped in clusters

D5 (*cont.*)   Grain faint
Rays distinct
Soft
Light

Peculiarities: Even, coppery-red colour, finely mottled. Pores
grouped in irregular bands. Sapwood yellowish,
clearly defined.

D6 MAHOGANY, HONDURAS

Group
features: One-coloured
Rings obscure
Pores diffuse, coarse
Grain faint
Rays distinct
Soft
Light

Peculiarities: Coppery-red shade, often broadly mottled with
darker or paler patches on an irregular plan.
Pores in irregular patches. Sapwood colourless,
well defined.

D7 PEARWOOD

Group
features: One-coloured
Rings obscure
Pores diffuse, very fine
Grain obscure or invisible
Rays obscure
Hard
Heavy

Peculiarities: Even red-brown colour without any apparent
features. Dull surface. Sapwood yellowish white.

Group features:	One-coloured, or broadly striped (see below)

Group
features:
One-coloured, or broadly striped (see below)
Rings distinct
Pores diffuse, coarse
Grain faint
Rays obscure
Hard
Heavy
Smell: aromatic

Peculiarities:
Coppery-red colour resembling mahoganies but distinguished by clearer rings, greater hardness and cedar-like smell. Often shows regular striped figure, with broad alternating pink and red-brown bands. Sapwood pale pink.

E: Wood Is brownish

E1 AFRORMOSIA

Group
features:
Two-coloured
Rings distinct
Pores diffuse, fine
Grain faint
Rays obscure
Hard
Heavy

Peculiarities:
Even striped pattern of golden brown alternating with warm crimson brown appears on radial surfaces. Sapwood yellow.

E2 CEDAR OF LEBANON

Group
features:
Two-coloured
Rings distinct

E 2 *(cont.)*   Pores absent
                Grain distinct
                Rays obscure
                Soft
                Very light
                Smell: aromatic, resin canals present

Peculiarities: Regular alternation of pale brown springwood
                and dark brown summerwood. Resin. Odour
                recalling incense. Sapwood whitish.

E3 CHERRY

Group   One-coloured, golden brown with a hint of green
features: Rings distinct
                Pores diffuse, fine
                Grain distinct
                Rays distinct
                Hard
                Light
                Smell: pleasant, faint, recalling rose blossom

Peculiarities: Mid-brown timber with a greenish-gold sheen
                and lively dappled figure. Sapwood pinkish.

E4 ELM

Group   One-coloured: warm brown
features: Rings distinct
                Pores diffuse, coarse to fine
                Grain distinct
                Rays distinct
                Soft
                Light
                Smell: earthy

Peculiarities: Irregular pores show on cross-section as zigzag
                lines between annual rings, and as wavy lines on
                longitudinal surfaces, giving 'partridge-breast'
                figure. Sapwood whitish.

Group features:	One-coloured: light brown Rings distinct Pores diffuse, coarse Grain distinct Rays obscure Hard Heavy
Peculiarities:	Resembles ash, but pores are scattered, not grouped in rings. Sapwood not defined.

E6 IROKO

Group features:	Two-coloured: brown with yellow mottling Rings distinct Pores diffuse, fine Grain distinct Rays obscure Hard Heavy
Peculiarities:	Resembles teak, but pores are diffuse, not in rings. Dull-surfaced. Sapwood not defined.

E7 OAK, BROWN

Group features:	One-coloured: deep brown Rings distinct Pores in rings, coarse Grain distinct Rays distinct Hard Heavy Smell: tannic acid
Peculiarities:	Resembles common oak, but heartwood is stained throughout to an attractive deep brown colour. Sapwood ill-defined.

E8 ROSEWOOD, BRAZILIAN

Group features:	Two-coloured Rings distinct Pores diffuse, coarse Grain distinct Rays obscure Hard Very heavy Smell: pleasant, recalling rose blossom
Peculiarities:	Dense, smooth timber with attractive odour, showing reddish-brown, golden-brown and violet-brown zones, enlivened by beautiful marbled figuring due to thin light and dark veins. Sapwood greyish.

E9 TEAK

Group features:	One-coloured: golden brown Rings distinct Pores in rings, coarse Grain distinct Rays obscure Hard Heavy Smell: leathery
Peculiarities:	Strong grain. Even colour. Pores in clear rings. Dull, rough surface. Oily feel. Sapwood yellow.

E10 WALNUT, AUSTRALIAN

Group features:	Two-coloured: pinkish-brown, with blackish-brown streaks Rings obscure Pores diffuse, coarse Grain faint Rays obscure Hard Heavy Smell: unpleasant

E 1 0 (*cont.*)

Peculiarities: Distinguished from American and European walnuts by ill-defined rings, wider range of colouring, greater weight and unpleasant smell. Sapwood pale brown.

E 1 1 WALNUT, BLACK AMERICAN

Group
features:
One-coloured: chocolate to purplish brown
Rings distinct
Pores diffuse, coarse
Grain faint
Rays obscure
Hard
Heavy
Smell: aromatic, mild

Peculiarities: Black American walnut is darker in colour than the Circassian kind. May be figured with lighter or darker streaks following annual rings. Distinctive odour. Sapwood pale yellow.

E 1 2 WALNUT, CIRCASSIAN

Group
features:
Two-coloured: greyish brown with darker streaks
Rings distinct
Pores diffuse, coarse
Grain faint
Rays distinct
Hard
Light

Peculiarities: The lively colour variations of Circassian walnut, due to dark brown, dark grey or black streaks following the ring pattern over a grey-brown ground, are highly distinctive. Rays visible. Sapwood pale yellow.

F: Wood is blackish or greyish

FI EBONY

Group
features:

Two-coloured (exceptionally one-coloured –
wholly black)
Rings obscure
Pores diffuse, fine
Grain obscure or invisible
Rays obscure
Extremely hard
Very heavy

Peculiarities:

Jet-black colour, streaked with brown or greenish black. Sapwood yellowish white or pinkish.

F2 PALDAO

Group
features:

Two-coloured
Rings obscure
Pores diffuse, coarse
Grain faint
Rays obscure
Hard
Very heavy

Peculiarities:

Highly figured wood showing dark brown,
greenish-brown or black streaks on a grey,
pinkish-grey or yellowish-grey ground. Resembles walnut. Sapwood pale pink.

KEY 2. SECONDARY COLOUR

Many woods show a single, consistent colour on their surfaces.
Where this occurs, the wood is termed 'one-coloured', and the
timbers concerned must be distinguished from one another by
other features. All other woods are here termed 'two-coloured'.
As the combinations of colour are rarely the same for any two
kinds, they provide valuable guides for identification. When using
this key the inquirer should not concern himself with the finer
structure that gives rise to the various colour combinations. All
he need note is their general effect on the exposed surface that he
sees. The secondary colour considered here is that found in the

heartwood of the timber, which is all that will be available on most specimens.

Sapwood colour. As explained earlier (p. 35), the outer layers of every growing tree and round log consist of *sapwood*. On all 'two-coloured' woods this is paler than the heartwood within, and this gives a further contrast. This feature, which is only likely to be found on round timber or broad planks, should be *ignored* when using Key 2. The sapwood shows up as such by its position as a *broad* pale band encircling a round log, or a *broad* irregular band on one side only of a piece of worked timber. A few decorative veneers, and some wide planks, are cut so as to show pale sapwood on two sides, but again it appears as broad bands. Very few 'one-coloured' woods show paler sapwood. As a rule no clear dividing-line appears between their heartwood and their sapwood.

Sapwood colour is noted in Key 1 under 'Peculiarities' for each timber. 'Sapwood not defined' implies that the general colour runs through both heartwood and sapwood. Woods with good sapwood definition are also listed in Key 13.

(a) One-coloured woods

Whitish
Ash A1
Aspen Poplar A2
Avodiré A3
Maple A5
Sycamore Maple A6

Reddish
Agba D1
Mahogany, African, D5
Mahogany, Honduras, D6
Pearwood D7
Sapele D8

Yellowish
Antiaris B1
Birch B2
Lacewood Plane B3
Lime B4
Primavera B7

Brownish
Cherry E3
Elm E4
Eucalyptus E5
Oak, Brown, E7
Teak, E9

Oak B5
Satinwood, East Indian, B8

Walnut, Black American, E11

Purplish or crimson
Purpleheart C2

(b) Two-coloured woods

Whitish
Brown spots and dark circles on white background: Bird's-eye Maple A4.

Yellowish

Pale yellow or golden-yellow rays (see Key 6) against brownish-yellow background, apparent only on radial surfaces: Oak B5.

Bright yellow springwood bands and red-brown summerwood bands in each annual ring; low contrast: Pine, Ponderosa, B6.

Vivid yellow springwood bands and dark brown summerwood bands following trend of rings; high contrast: Zebrawood B9.

Purplish or Crimson

Black or dark purplish-brown streaks on crimson or purplish ground: Padouk, Andaman, C1.

Even alternation of purple or crimson stripes on red-brown ground: Bubinga C3.

Reddish

Small, regular, chocolate-brown flecks or plates on pinkish-brown ground: Beech D2.

Red-brown summerwood and yellowish-pink springwood in each annual ring; high contrast: Douglas Fir D4.

Red-brown summerwood and yellow-brown springwood in each annual ring; low contrast: Cedar, Western Red, D3.

Broad pink bands alternating with broad red-brown bands: Sapele D8.

Brownish

Golden-brown bands alternating with warm brown bands in each ring: Afrormosia E1.

Pale brown springwood and dark brown summerwood bands alternating: Cedar of Lebanon E2.

Yellow stripes on brown ground: Iroko E6.

Reddish-brown with both paler and darker streaks: Rosewood, Brazilian, E8

Brown, striped with grey, green, black or pink: Walnut, Australian, E10.

Greyish brown, with darker zones, and irregular dark brown to black streaks marking outer edge of each ring: Walnut, Circassian E12.

Blackish or Greyish

Grey or yellowish-grey ground bearing dark brown or black streaks: Paldao F2.

Black ground striped with bright brown or greenish black: Ebony F1.

The annual rings or growth rings that run through every piece of wood are the next feature to note. On a cross-cut or transverse surface visible rings always show a curved pattern. If the piece of wood concerned was cut near the heart of the tree, the curve will be quite sharp and circular. If it was cut farther out the curvature will be less, yet still apparent. It will be clear from the account of how wood is formed (p. 30) that complete rings always form circles, but as a rule you see only part of each circle.

On longitudinal surfaces the rings show as bands that run for virtually the whole length of the piece of wood. If the surface is parallel to the axis of the stem they will in fact run the whole length. Otherwise certain rings may run out to the side, or to the edge, part-way along. This feature of consistent length distinguishes rings from rays, which are considered later, in Key 6.

Those woods that show annual rings plainly are listed here under the heading 'Rings distinct'.

Certain trees that grow in tropical countries with no variation in climate over the year yield timbers that lack apparent rings. A few temperate-zone trees have woods of a colour and consistency that make their rings hard to see, though present. These two groups are listed together in this key under the second classification, 'Rings obscure'.

(a) Rings distinct

Ash A1	Douglas Fir D4
Bird's-Eye Maple A4	Sapele D8
Maple A5	Afrormosia E1
Lacewood Plane B3	Cedar of Lebanon E2
Teak E9	Cherry E3
Oak B5	Elm E4
Pine, Ponderosa, B6	Eucalyptus E5
Primavera B7	Iroko E6
Zebrawood B9	Oak, Brown, E7
Padouk, Andaman C1	Rosewood, Brazilian, E8
Bubinga C3	Walnut, Black American, E11
Agba D1	Walnut, Circassian, E12
Cedar, Western Red, D3	

(b) Rings obscure

Aspen Poplar A2
Avodiré A3
Sycamore Maple A6
Antiaris B1
Birch B2

Lime B4
Satinwood, East Indian, B8
Purpleheart C2

Beech D2
Mahogany, African, D5

Mahogany, Honduras, D6
Pearwood D7
Walnut, Australian, E10
Ebony F1
Paldao F2

KEY 4. PORES

Associated with the rings in precisely the same circular pattern
come the pores, or vessels. On the cross-cut surface they show as
small circular holes usually visible to the naked eye, and always
apparent – if present – under a hand lens. On longitudinal surfaces
they form shallow grooves or striations that run, like the rings,
for virtually the whole length of the piece of wood examined.

There are three possibilities here, which divide all timbers into
three convenient groups:

 (a) *Pores absent*. Softwoods, yielded by coniferous trees, which *never* develop pores.

 (b) *Pores in rings*. Certain hardwoods form
large pores, in distinct circles, at the inner
edge of each annual ring.

 (c) *Pores diffuse*. Other hardwoods have
small pores scattered evenly right through
each ring.

In some diffuse-porous woods all the pores are fine. In others, and
in *all* the ring-porous woods, there is a proportion of large pores.
These are best seen as coarse striations or vessel lines running
along longitudinal surfaces. So a further useful distinction can be
made between woods showing *coarse* pores and those with *fine*
pores, hard to detect on longitudinal surfaces.

(a) Pores absent

All coniferous trees or softwoods, including:
Pine, Ponderosa, B6
Cedar, Western Red, D3
Douglas Fir D4
Cedar of Lebanon E2

(b) Pores in rings

Pores form obvious rings associated with the springwood of the annual rings. Always *coarse*.
Ash A1
Oak B5
Padouk, Andaman, C1
Oak, Brown, E7
Teak E9

(c) Pores diffuse

Pores scattered through the width of the annual rings. On longitudinal surfaces they may appear as vessel lines that are fine and barely visible, or as ones that are coarse.

Fine pores and lines	*Coarse pores and lines*
Aspen Poplar A2	Antiaris B1
Avodiré A3	Primavera B7
Bird's-eye Maple A4	Zebrawood B9
Maple A5	Agba D1
Sycamore Maple A6	Mahogany, African, D5
Birch B2	Mahogany, Honduras, D6
Lacewood Plane B3	Sapele D8
Lime B4	Elm E4
Satinwood, East Indian, B8	Eucalyptus E5
Purpleheart C2	Rosewood, Brazilian, E8
Bubinga C3	Walnut,
Beech D2	Australian, E10
Pearwood D7	Walnut,
Afrormosia E1	Black American, E11
Cherry E3	Walnut,
	Circassian, E12
Iroko E6	
Ebony F1	Paldao F2

The rings and the pores together produce a general visual effect on all longitudinal surfaces that is known as the 'grain'. Differences in consistency, size of wood cells and colour result in three broad but useful classifications: *Grain distinct*; *Grain faint* (grain appears as faint lines but is quite discernible); *Grain obscure or invisible* (no apparent grain). Rays, which are considered in Key 6, also contribute to the character of each timber's grain.

When used in the strict sense, the word 'grain' implies the physical texture of the wood, which governs its working properties. But it is also the most convenient term for the visual effect of that texture.

(a) Grain distinct

Ash A1	Douglas Fir D4
Bird's-Eye Maple A4	
Maple A5	Eucalyptus E5
Oak B5	Cedar of Lebanon E2
Teak E9	Cherry E3
	Elm E4
Pine, Ponderosa, B6	Iroko E6
Primavera B7	
Zebrawood B9	Oak, Brown, E7
Padouk, Andaman, C1	Rosewood, Brazilian, E8

(b) Grain faint

Avodiré A3	Mahogany, African, D5
Sycamore Maple A6	
Birch B2	Mahogany, Honduras, D6
Lacewood Plane B3	Sapele D8
Purpleheart C2	Afrormosia E1
	Walnut, Australian, E10
Bubinga C3	Walnut, Black American, E11
Agba D1	
Beech D2	Walnut, Circassian, E12
Cedar, Western Red, D3	Paldao F2

(c) Grain obscure or invisible

Aspen Poplar A2	Satinwood, East Indian, B8
Antiaris B1	Pearwood D7
Lime B4	Ebony F1

Rays are found in all timbers, and may be conspicuous or so fine that they can barely be seen even with a hand lens. On cross-cut surfaces they show as narrow solid bands radiating out from the tree's centre. Few pieces of wood include this actual centre, but rays can always be distinguished as 'straight-line' tissues cutting straight across the curved rings.

On longitudinal surfaces rays show as flat plates or thin bands, often bright and shiny, that are relatively shallow. They can be told apart from the rings because they *never* run consistently through the whole length of any reasonably sized sample; in fact, except in oak, they are seldom 'deeper' than a fraction of an inch. On surfaces cut radially – from the centre of the tree towards its edge – rays are often conspicuous by reason of peculiar colour, texture and hardness, and also highly decorative. On other longitudinal surfaces they are narrow.

All surfaces of each piece of wood should therefore be examined for rays. After this, it is easy to split timbers into two groups: *Rays distinct* and *Rays obscure*. The visual effect of the rays is known as 'ray figure'.

(a) Rays distinct

Bird's-eye Maple A4	Beech D2
Maple A5	Mahogany, African, D5
Sycamore Maple A6	Mahogany, Honduras, D6
Lacewood Plane B3	Cherry E3
Oak B5	Elm E4
Primavera B7	Oak, Brown, E7
Bubinga C3	Walnut, Circassian, E12

(b) Rays obscure

Ash A1	Satinwood, East Indian, B8
Aspen Poplar A2	Zebrawood B9
Avodiré A3	Padouk, Andaman, C1
Antiaris B1	Purpleheart C2
Birch B2	Agba D1
Lime B4	Cedar, Western Red, D3
Pine, Ponderosa, B6	Douglas Fir D4

Pearwood D7   Rosewood, Brazilian, E8
Sapele D8   Teak E9
Afrormosia E1   Walnut, Australian, E10
Cedar of Lebanon E2   Walnut, Black American, E11
Eucalyptus E5   Ebony F1
Iroko E6   Paldao F2

KEY 7. HARDNESS

Woods vary a great deal in their surface hardness or resistance to indentation. This can easily be assessed by trying to press a thumb nail or the point of a ball-point pen into the specimen. When so doing, avoid any marked band of dark summerwood, which will be harder than average. Two groups can be distinguished in this way: *soft*, easily indented, and *hard*, resisting indentation. Very hard or soft woods are specially noted later.

The 'hardness' considered here has nothing to do with the conventional trade definition whereby all timber from broad-leaved trees is termed 'hardwood', whether it is actually hard or soft, or the parallel trade term of 'softwood', which covers the timbers of all conifers.

(a) Soft

Aspen Poplar A2   Douglas Fir D4
Antiaris B1   Mahogany, African, D5
Lime B4   Mahogany, Honduras, D6
Pine, Ponderosa, B6   Cedar of Lebanon E2
Agba D1   Elm E4
Cedar, Western Red, D3

(b) Hard

Ash A1   Zebrawood B9
Avodiré A3   Padouk, Andaman, C1
Bird's-eye Maple A4   Purpleheart C2
Maple A5   Bubinga C3
Sycamore Maple A6   Beech D2

Birch B2   Pearwood D7
Lacewood Plane B3   Sapele D8
Oak B5   Afrormosia E1
Primavera B7   Cherry E3
Satinwood, East Indian, B8   Eucalyptus E5

(b) *(cont.)*

Iroko E6

Oak, Brown, E7

Rosewood, Brazilian, E8

Teak E9

Walnut, Australian, E10

Walnut, Black American, E11

Walnut, Circassian, E12

Ebony F1

Paldao F2

Exceptionally hard woods are Bird's-eye Maple A4, Maple A5 and Ebony F1. Exceptionally soft are Aspen Poplar A2 and Antiaris B1.

KEY 8. WEIGHT

Weight here implies the general density or specific gravity of each piece of wood. Wood is made up of solid cell walls which enclose space. When a tree is growing this space is full of sap, and there is very little difference between the specific gravity of one kind of tree and that of another. The same is true of freshly felled logs and 'green' timber generally.

During the process of seasoning nearly all the water disappears, and the density of the timber then depends mainly on two things: the thickness of the cell walls, and the amount of empty space between them. Timbers vary a great deal in their fine structure, some holding *thick* cell walls with *small* spaces, others having *thin* cell walls with *large* spaces. The density of seasoned, that is, air-dried, timbers can therefore be a great help towards their identification. Some weigh four times as much as others, for the same volume.

Two scales of measurement are in use for wood density, so both are given here. One, called *specific gravity* (*S.G.*), represents the ratio of each wood's density to that of water, which has a specific gravity of 1·0. Since most seasoned woods are lighter than water, their specific gravity is a decimal fraction; it also shows the number of grams in one cubic centimetre. The alternative scale shows the weight of one cubic foot of timber, that is, so many pounds per cubic foot. (One cubic foot of water weighs $62\frac{1}{2}$ lb.)

The following key shows average densities for forty woods. There are a few borderline cases, but it will be seen that most woods fall clearly into four main groups:

Very light: below 0·48 S.G. or 30 lb./cu.ft.

Light: between 0·48 S.G. or 30 lb./cu.ft. and 0·64 S.G. or 40 lb./cu.ft.

Heavy: between 0·64 S.G. or 40 lb./cu.ft. and 0·80 S.G. or 50 lb./cu.ft.

Very heavy: over 0·80 S.G. or 50 lb./cu.ft.

These differences are so marked that the 'feel' of a piece of wood often places it within its class. If confirmation is needed, it is a simple matter to weigh it, ascertain its volume and calculate its actual density.

Average densities of forty woods

Wood				S.G.	lb./cu. ft.
VERY LIGHT					
Aspen Poplar A2	..	..	..	0·44	27
Antiaris B1	..	..	..	0·44	27
Primavera B7	..	..	..	0·46	29
Cedar, Western Red, D3	..	..	..	0·38	23
Cedar of Lebanon E2		..	..	0·45	28
LIGHT					
Avodiré A3	..		..	0·58	36
Sycamore Maple A6	..		..	0·56	35
Lacewood Plane B3	..		..	0·63	39
Lime B4	..	..	..	0·56	35
Pine, Ponderosa, B6	..		..	0·51	32
Agba D1	..	..	..	0·50	31
Douglas Fir D4	..	..	..	0·50	31
Mahogany, African, D5	..		..	0·50	31
Mahogany, Honduras, D6		..	..	0·50	31
Cherry E3	..	..	..	0·63	39
Elm E4	..	..	..	0·51	32
Walnut, Circassian, E12	..		..	0·63	39
HEAVY					
Ash A1	..	..	..	0·69	43
Bird's-Eye Maple A4	..		..	0·69	43
Maple A5	..	..	..	0·69	43
Birch B2	..	..	..	0·67	42
Oak B5	..	..	..	0·69	43

Zebrawood B9	..	..	..	0·68	42
Beech D2	..	..	..	..	0·67	42
Pearwood D7	..	..	..	0·69	43
Sapele D8	..	..	..	..	0·67	42
Afrormosia E1	..	..	..	0·74	46
Eucalyptus E5	..	..	..	0·67	42
Iroko E6	..	..	..	..	0·65	41
Oak, Brown, E7	..	..	..	0·69	43
Teak E9	..	..	..	..	0·65	41
Walnut, Australian, E10	..	..	0·77	48
Walnut, Black American, E11	..	0·66	42

VERY HEAVY

Satinwood, East Indian, B8	..	0·88	55
Padouk, Andaman, C1	..	..	0·83	52
Purpleheart C2	..	..	..	0·86	54
Bubinga C3	..	..	..	0·91	57
Rosewood, Brazilian, E8	..	..	0·88	55
Ebony F1	..	..	..	..	1·08	63
Paldao F2	..	..	..	..	0·83	52

KEY 9. SMELL

Certain woods hold volatile chemicals, usually ethereal oils, that gradually escape when their cells are cut across, so yielding characteristic odours of great value for identification. Freshly felled and freshly cut timber has naturally the strongest odour. As the wood ages its smell decreases, but it can often be revived, even after many years, by exposing a fresh surface to the air. Sometimes the scent is sealed in by varnish; again exposure will release it.

In most conifers scent is due to resin, which is often contained in special resin canals that run through the wood. Where these occur, their presence is noted in this key. Resin often exudes on cut surfaces as a clear yellow, sticky substance with a characteristic 'turpentine' smell. Only timbers with a characteristic smell are listed here:

(a) *Peppery smell:*	Agba D1
(b) *Unpleasant smell:*	Antiaris B1
	Zebrawood B9
	Walnut, Australian, E10
(c) *Pleasant smell:*	Lime B4
	Padouk, Andaman, C1
	Cherry E3
	Rosewood, Brazilian, E8
(d) *Sharp, tannic-acid smell:*	Oak B5
	Oak, Brown, E7
(e) *Resinous smell:*	Pine, Ponderosa, B6
	Douglas Fir D4
(f) *Earthy smell:*	Elm E4
(g) *Aromatic smell:*	Cedar, Western Red, D3
	Sapele D8
	Cedar of Lebanon E2
	Walnut, Black American, E11
(h) *Leathery smell:*	Teak E9

KEY 10. BARK

Tropical woods are omitted from this key because they are imported as round logs or sawn baulks from which all bark has been removed. On young, thin stems, bark is always fairly smooth and featureless, but as tree-trunks enlarge to sawmill size they develop bark patterns that are highly characteristic for each kind. These patterns, though easy to pick out through observation, are hard to define in words. Here we distinguish six categories, listed from (a) to (f) below.

(a) *Bark smooth, firm, thin:*	Aspen Poplar A2
	Birch B2
	Beech D2
	Cherry E3

(b) *Bark smoothish, breaking away in round, flat plates:*	Sycamore Maple A6 Lacewood Plane B3 Eucalyptus E5
(c) *Bark smoothish, strongly fibrous:*	Lime B4 Cedar, Western Red, D3
(d) *Bark thick, furrowed, with loose, fibrous plates:*	Pine, Ponderosa, B6
(e) *Bark moderately thick, with shallow furrows, resinous:*	Douglas Fir D4

(f) *Bark thick, ridged and furrowed, firm:*

Ash A1	Cedar of Lebanon E2
Bird's-eye Maple A4	Elm E4
Maple A5	Oak, Brown, E7
Oak B5	Walnut, Black American, E11
Pearwood D7	Walnut, Circassian, E12

KEY 11. LEAF SHAPE

Leaf shape is constant for each kind of tree, and wherever leaves are available they give a handy clue to each timber's identity. But leaves are usually left behind in the forest, sometimes many thousands of miles away. Only those leaves that are likely to be seen in temperate countries are included here; many of them are illustrated by sketches in Part III. The main classifications for leaves are as follows:

(a) *Simple*: An undivided leaf-blade, oval or round in outline, with only one main vein.

(b) *Lobed*: An undivided leaf-blade that has several distinct lobes, each holding its own main vein running out to a point.

(c) *Compound*: Leaf divided into many separate small leaflets. (Compound leaves are easily distinguished from *shoots* because they never bear buds.)

(d) *Needle-shaped*: Very slender, undivided. (All the trees in this group are conifers or softwood trees.)

(a) Leaf simple

Aspen Poplar A2   Pearwood D7
Birch B2   Cherry E3
Lime B4   Elm E4
Beech D2   Eucalyptus E5

(b) Leaf lobed

Bird's-eye Maple A4   Lacewood Plane B3
Maple A5   Oak B5
Sycamore Maple A6   Oak, Brown, E7

(c) Leaf compound

Ash A1
Walnut, Black American, E11
Walnut, Circassian, E12

(d) Leaf needle-shaped

Pine, Ponderosa, B6   Douglas Fir D4
Cedar, Western Red, D3   Cedar of Lebanon E2

KEY 12. COUNTRY OF ORIGIN

If the country of origin of a piece of wood is known, it may provide a useful short-cut to the determination of its identity. At the least it will save your considering several possible trees that cannot possibly apply, because the wood concerned never grows in their homelands. But this key must be applied with caution because woods can travel a long way in the course of international commerce. For example, a log that grew in Brazil may cross the Atlantic to Italy to be cut into veneer, and then recross the ocean again to be applied to furniture made in the United States. Still, it is a great help to be able to eliminate, say, the tropical hardwoods when you have to name a piece of wood known to have originated in some temperate land.

This key is, of necessity, a generalized one, because several common timbers occur naturally, or have been planted, in several parts of the temperate zone. For example, oaks similar to the American species grow wild in Europe and Japan, while the Douglas fir, native only to North America, has been extensively planted both in Europe and Australia.

(a) North America

Ash A1
Aspen Poplar A2
Bird's-eye Maple A4
Maple A5
Birch B2

Lacewood Plane B3
Lime B4
Oak B5
Pine, Ponderosa, B6
Beech D2

Cedar, Western Red, D3
Douglas Fir D4
Pearwood D7
Cherry E3
Elm E4

Walnut, Black American, E11

(b) Europe

Ash A1
Aspen Poplar A2
Sycamore Maple A6
Birch B2
Lacewood Plane B3

Lime B4
Oak B5
Beech D2
Cedar, Western Red, D3
Douglas Fir D4

Pearwood D7
Cedar of Lebanon E2
Cherry E3
Elm E4
Oak, Brown, E7

Walnut, Circassian, E12

(c) Northern Asia and Japan

Ash A1
Aspen Poplar A2
Birch B2
Oak B5
Beech D2

Elm E4

(d) Australia

Douglas Fir D4
Eucalyptus E5
Walnut, Australian, E10

(e) Tropical America

Primavera B7
Purpleheart C2
Mahogany, Honduras, D6
Rosewood, Brazilian, E8

(f) Tropical Africa

Avodiré A3
Antiaris B1
Zebrawood B9
Bubinga C3
Agba D1

Mahogany, African, D5
Sapele D8
Afrormosia E1
Iroko E6

(g) Tropical Asia and its islands

Satinwood, East Indian, B8
Padouk, Andaman, C1
Teak E9
Ebony F1
Paldao F2

74

If a specimen shows a clear distinction between darker heartwood and paler sapwood, both in broad bands, a useful clue is provided to its possible identity. The absence of such definition, however, does not rule out *any* timber, because many specimens show only heartwood. *Sapwood can be well defined in:*

Oak B5	Bubinga C3
Pine, Ponderosa, B6	Cedar, Western Red, D3
Zebrawood B9	Douglas Fir D4
Padouk, Andaman, C1	Mahogany, African, D5
Purpleheart C2	Mahogany, Honduras, D6
Pearwood D7	Rosewood, Brazilian, E8
Sapele D8	Teak E9
Afrormosia E1	Walnut, Australian, E10
Cedar of Lebanon E2	Walnut, Black American, E11
Cherry E3	Walnut, Circassian, E12
Elm E4	Paldao F2
Ebony F1	

KEY 14. CLASS OF USE

Ideally, you should be able to identify a timber quite regardless of any past or probable future use. In practice, it is a handy short-cut to discover the purpose that a specimen has already served, or may serve in the future.

This will, for example, save you a waste of time in considering whether a yellowish piece of fence post may possibly be the precious East Indian Satinwood, used only indoors as veneer or in high-grade decorative woodwork. It is more likely to be oak, which is far cheaper and far more readily available in the size needed.

(a) Precious woods
Used for the finest decorative wood-carving, highest grade furniture and cabinet-work or as veneer.

Bird's-eye Maple A4	Zebrawood B9
Satinwood, East Indian, B8	Purpleheart C2

75

Bubinga C3	Walnut, Australian, E10
Pearwood D7	Walnut, Circassian, E12
Oak, Brown, E7	Ebony F1
Rosewood, Brazilian, E8	Paldao F2

(b) General-purpose furniture woods

Used 'in the solid' for the general framework of large pieces of furniture, and in joinery, etc., mainly indoors.

Ash A1	Antiaris B1
Aspen Poplar A2	Birch B2
Avodiré A3	Lacewood Plane B3
Maple A5	Lime B4
Sycamore Maple A6	Oak B5

Pine, Ponderosa, B6	Cedar, Western Red, D3
Primavera B7	Douglas Fir D4
Padouk, Andaman, C1	Mahogany, African, D5
Agba D1	Mahogany, Honduras, D6
Beech D2	Sapele D8

Afrormosia E1	Eucalyptus E5
Cedar of Lebanon E2	Iroko E6
Cherry E3	Teak E9
Elm E4	Walnut, Black American, E11

(c) Everyday woods

Applied to many kinds of ordinary construction and household objects, indoors and out, because of low cost and ready availability.

Aspen Poplar A2	Beech D2
Maple A5	Cedar, Western Red, D3
Birch B2	Douglas Fir D4
Oak B5	Elm E4
Pine, Ponderosa, B6	

5 Examples of the Use of Keys

First, a few general hints on how to look at a piece of wood presented for identification.

Is it really wood? Make sure at the outset that it *is* wood. The surface of timber is so attractive that it is often simulated in other materials. Even an expert may be fooled, in a bad light, by his first glance at a good reproduction made by a modern photographic technique. Even solid oak beams have been faked with fibre-glass!

The usual materials for the imitation of real wood surfaces are wallpaper and plastic laminates. Both are betrayed by the lifeless-ness of their surface, the lack of depth to the texture, and the precise repetition of the same pattern at frequent intervals. They are pictures of wood, like the pictures in a book. They lack the irregularity of the real material, and their colours are those of the printer's ink works, not of the natural laboratory of the woods.

Occasionally you may encounter 'densified wood' which has been made harder and heavier by impregnation with plastic resin, under heat and pressure. But this is still wood, and holds its original structure.

Veneer or solid? Your next problem, particularly when examining furniture, is to decide whether you are dealing with solid wood, or a surface veneer. Decorative veneers are often skilfully applied to curved surfaces and edges, concealing the commoner wood that lies below. But, on grounds of cost, it is almost unknown for *all* the surfaces of a wooden object to be veneered. So look for the faces and edges that 'don't show' when the specimen is in ordinary everyday use.

Obviously you will not expect to find cheap or common objects, such as tool-handles, hidden by veneer. But furniture, ornamental boxes, radio cabinets, trays and a whole range of more expensive woodware are often veneered.

The run of the grain. The next move, with either solid wood or a surface veneer, is to decide the main direction of the wood's grain, which corresponds to the main axis of the trunk of the tree in which it grew. Every piece of wood has this feature, no matter how it has been manipulated since the tree that produced it was felled.

A solid object should be moved around until you have found a cross-section showing its annual rings as circles and its rays making a star-shaped pattern. Even if both these features are obscure, the roughness of the end-grain is usually apparent.

With the wood orientated in this way, you can detect the two kinds of longitudinal surface that show on its sides – the radial surfaces following the plane of the rays, and the tangential surfaces that follow – though less precisely – the outer faces of the annual rings.

On veneered furniture, sets of matched veneers are often grouped together with their grain running in different directions to create a pattern. Once this is realized, it should cause no problems.

Veneers scarcely show end-grain at all, but most of them will reveal *either* ray surfaces *or* tangential surfaces very clearly. A few, owing to oblique cutting, may prove difficult, but general structure is always apparent.

Consider all the evidence. No two surfaces of a piece of wood, nor even two parts of the same surface, are ever precisely alike. Clues to identity may not show everywhere, but they will always appear *somewhere*.

Example 1

You are asked to identify a square piece of wood, measuring one foot each way by half an inch thick, which has served as a table-top. One side only has been coated with clear varnish. It is clearly solid, not veneered, and the grain runs straight across from one end to another.

Taking Key 1, the Main Key, first, you consider the wood's general colour. It is bright yellow on the varnished side, dull yellow elsewhere. You consider other possibilities in turn, whitish, reddish, purplish, brownish and blackish, and decide that none really applies. You therefore decide that your wood does in fact lie in Group B, *yellowish* wood.

Now take Key 2, Secondary Colour. Is the yellowish wood one-coloured or two-coloured? The unvarnished surfaces appear one-coloured, but the varnished surface shows two distinct shades of yellow – bright and dark. So it *might* be oak (occasionally two-coloured), or perhaps another one-coloured yellow wood. You look for sapwood, which might show as a broad pale band, but can find none, so Key 13, Sapwood Definition, cannot help your search.

Next you try Key 3, for growth rings. You look at both the sides and the ends of your piece of wood and discover that two opposite ends show a pattern of curved lines. Rings are therefore *distinct*, and you can eliminate from your choice all yellowish woods that have obscure rings. Turn back now to the Main Key and look through all the yellowish woods in Group B. You will see that antiaris, lime and East Indian satinwood are all ruled out, but your wood might still be any one of the following: birch, lacewood plane, oak, ponderosa pine, primavera or zebrawood.

Next take Key 4, Pores. Another look at those ends of the piece of wood where the rings appeared reveals clear tiny holes, in narrow bands following the course of the rings. Pores are therefore present and you can rule out all woods that lack them. Further, they are in *rings*, and you can narrow your search down to those woods shown in that group in Key 5. There are only five of these, and only one, namely *oak*, is yellowish. All the other yellowish woods with distinct rings have diffuse pores, or else no pores at all.

As a further check, look at the flat unvarnished side of your piece of wood. It shows frequent deep striations running through its whole length: these are the vessel lines that represent the side view of pores. You can therefore describe your pores as *coarse*. Oak fits in here, and another look at the Main Key will show that it has become the only choice. No other yellowish wood has distinct rings, and also coarse pores following those rings.

But to make quite sure, try as many other keys as possible. Key 5, Grain on Longitudinal Surfaces, confirms oak, for your specimen shows pronounced grain. Turning to Key 6, you look for rays, and find they appear as broad pale bands on the varnished surface, giving it the two-coloured appearance you noted earlier. Since they are not very deep, each running for only half an inch or so down the surface, you can be sure they *are* rays and not something else. Distinct rays help to confirm oak.

Next you test hardness, Key 7, and discover that you cannot dent the surface of this wood with your thumb nail or a ball-point pen. So it ranks as *hard*, which again covers oak. Even if it is *not* oak, you have eliminated all the soft woods by this simple test.

Now for its *weight*, or rather its specific gravity or density in pounds per cubic foot. Since this wood has been well seasoned through years of use indoors as a table-top, you can safely apply

Key 8. Your piece measures one foot square by half an inch thick, and from this you calculate its volume as one twenty-fourth part of a cubic foot. You put it on your scales and find it weighs $1\frac{3}{4}$ lb. Multiply this by twenty-four, and its weight per cubic foot must be 42 lb. This puts it in the class of *heavy* woods, and this again confirms oak. By this means you have ruled out all the Very light, Light, and Very heavy woods shown in Key 8.

You cannot apply Key 9 here, for your specimen is so old that it has lost all its natural smell. Key 10 is no help, since it carries no bark, nor is Key 11, for no leaves are present. Key 12, Country of Origin, cannot help, for a piece of wood like this may have originated anywhere; Key 13 does not help either, for the specimen shows no sapwood. But you gain·a little information from Key 14, Class of Use, which includes oak among those woods that are often used in the solid form for furniture.

You have still to take the critical step of comparing your piece of wood with the actual specimen at the front of this book. If you have run it down correctly you will be struck by the similarity – not merely in one feature but in many – colour, grain, pores, rays, rings and hardness. Do not expect a perfect match, since no two pieces of wood are ever exactly alike but merely display a family resemblance in all critical features.

Finally, turn to Part III of the book, and look up *oak*. There its general description will confirm your diagnosis and give you further details about this fascinating timber.

Example 2

A friend brings in from a distant forest a freshly cut, short, stout, round log and asks you to name it. Along with the log he brings a spray of foliage that *might* have come from the same tree – he is not sure. You look at the leaves and turn to Key 11, Leaf Shape. The leaves are so narrow that you can only place them in Group (*d*): Needle-shaped.

There are four possibilities here, all conifers or softwood trees, but you cannot say which, nor be sure that the leaves are linked with this log. But you consider, provisionally, that this tree may be a conifer. The run of the grain is clear – from top to bottom of the log, following the axis of the tree-trunk.

This log grew in a northern forest, so you know its country of origin. You turn to Key 12, which covers this, and you can rule out a large number of timbers that grow only in tropical lands. This narrows your field of search considerably.

Next you look at the wood on the exposed ends of the log. You see there a sharp colour distinction between dark wood at the centre and pale wood forming a ring round it. Key 13, Sapwood Definition, lists all those woods that show clearly marked sapwood, so your log may belong to one of the kinds named. Timbers that never show a well-marked colour range between heartwood and sapwood are all ruled out.

Bark is still present all round the log, so you consider Key 10. This bark is neither smooth nor fibrous, and you cannot describe it as deeply ridged or furrowed, nor as 'flaking away'. Its surface is broken by *shallow* furrows, and you suddenly notice a peculiar odour, resembling turpentine. You scratch the bark and this scent grows stronger, and you notice tiny beads of a clear yellow substance – resin. The key offers one possibility here – Douglas fir, D4, so you make this a provisional choice.

Turn now to the Main Key, where Douglas fir is shown at D4, to see how this identification fits. So far you have scarcely considered the actual timber, so there are several features to check.

First, you confirm that the general colour of the heartwood— that is, the zone within the paler sapwood – is reddish. Next, still concentrating your attention on this heartwood, you decide that it can be described as two-coloured (Key 2). There are clear circular bands of dark red-brown wood alternating with circular bands of pale yellowish-pink wood. This gives the answer also to the next question: Are the rings distinct (Key 3)? They are clearly so, and the name 'Douglas fir' still applies.

You search next for pores (Key 4), by looking hard, preferably with the aid of a hand lens, over the cut surface at one end. No pores appear, so you decide that your log falls in the class of softwoods, 4(a), which includes Douglas fir.

The next feature is grain on longitudinal surfaces (Key 5), but so far you have seen no such surface. So you take a saw and cut down through the log, making your cut close to its centre. The surfaces you have now exposed show that the grain is distinct, owing to strongly alternating colour and texture of summerwood and springwood. You then search for rays (Key 6), but no obvious rays appear; so on both counts your wood may still be Douglas fir.

Now test for hardness (Key 7), by pressing your thumb nail into one of the paler springwood bands. It goes in readily, so you rank this wood under 7(a), that is, soft. Again the name Douglas fir can still apply.

You cannot use Key 8, Weight, because your log is freshly cut and unseasoned. It holds so much sap that you cannot safely determine its density. Nor does Key 14, Class of Use, help, because you do not know how this raw log would eventually have been used. But Key 9, Smell, confirms the name Douglas fir, for the freshly sawn surface of your log has a marked resinous odour.

So everything points to *Douglas fir*, and you confirm this by comparing a longitudinal surface on your sawn log with the actual specimen included in this book. Finally you turn to the description of Douglas fir in Part III to check on all its properties and learn more about its occurrence and uses.

Example 3

You have bought a small antique carved wooden figure representing a partridge, and wish to know what wood the sculptor used. The wood has a distinct deep red-brown hue, and Key 1, General Colour, leads you to classify it in Group D, Wood is reddish.

You proceed to Key 2, Secondary Colour, and since the colour is even throughout, place your specimen in Group (*a*), the one-coloured woods. There are five of these in the 'Reddish' section, namely agba, African mahogany, Honduras mahogany, pearwood, and sapele.

Go on now to Key 3, and examine your specimen for rings. You turn it one way and another, but no rings are apparent. So you rank it in Group (*b*), Rings obscure. This rules out two of your previous selection, for it *cannot* be agba or sapele. Both those have distinct rings.

You are left with the choice of the two mahoganies and pearwood. You seek next for pores in Key 4. At first you see none, but close examination of the base of your model reveals very fine scattered holes. So you class your wood under 4(*c*), Pores diffuse. Are these pores fine or coarse? You try to trace them as vessel lines down the sides of your model, but they do not appear. Hence you rank them as *fine*, and a quick run down the key shows that this eliminates the two mahoganies – since both show coarse pores.

You are left with only one possibility, *pearwood*, so you turn to the point indicated, D7, in the Main Key, for a check.

This requires you to consider whether the grain and the rays are obscure, and further examination of your little carved bird confirms this. You test its hardness with your thumb nail, and agree that it is truly hard. You feel its weight in your hand, and decide that you can call it heavy. The peculiarities listed confirm that pearwood is featureless – apart from its remarkable colour.

The remaining keys are of little help: for the wood is odourless, you have no bark or leaf, and you cannot say where the wood originated. Nor is any visible sapwood present. But Key 14, Class of Use, confirms that pearwood is applied to decorative wood-carving.

You end your search by comparing your carving with the actual wood specimen in this book, and reading the general description in Part III. In effect you have 'run down' a fairly featureless wood by eliminating other similar kinds with well-marked features. On reflection, it was just this character that attracted the sculptor. He wished to express a pure bird shape without the conflicting impressions caused by wood figure and grain.

Example 4

A cocktail cabinet is gaily patterned in yellow and brown on all easily seen surfaces. You look at the hidden edge of a hinged shelf and discover that this is a dull, featureless brown. You then suspect that the whole cabinet is veneered, and confirm this by finding the thin edges of the veneer exposed at the corners of shelves and drawers.

Key 14, Class of Use, gives you a lead to twelve precious woods likely to be used as decorative veneer in an expensive piece of furniture, so you decide to run through these next.

By now you are familiar with the general run of the keys, and you realize that, as explained at the start of Key 1, woods prefixed 'A' are whitish, those prefixed 'C' are purplish or crimson, those prefixed 'D' are reddish or pinkish, and those prefixed 'F' are blackish or greyish. This eliminates six possibilities from the precious woods in Key 14. You still have six choices left, however, two appearing under 'B' as yellowish, four under 'E' as brownish.

Because the wood is so definitely – but puzzlingly – two-coloured, you turn to Key 2, Secondary Colour. Here you find that one possibility, B8, East Indian satinwood, must be dropped because it is one-coloured, being *all* yellow. Another, brown oak, likewise falls, because it is also one-coloured, but *all* brown. Three other choices, Brazilian rosewood, Australian walnut and Circassian walnut, are dropped because yellow is not apparent in their colour schemes.

Only one probability remains, *zebrawood*, B9, a two-coloured wood with 'striking contrast between yellow and brown bands following trend of annual rings'.

Your search, however, has brought in other possibilities. Ponderosa pine, B6, also appears in Key 2 as a light and dark yellowish wood, though it is not a 'precious' one. Could this be the name? You turn to the Main Key, Key 1, for some clear point of difference, and find it under 'pores'. These are described as 'absent' for the pine, but 'diffuse and coarse' for zebrawood. So you look at the specimen again and find the long striations or vessel lines that represent pores running down the face of your veneer. It cannot be pine, but zebrawood still fits.

Another two-coloured yellowish wood shown in Key 2 is oak. Again you seek points of difference in Key 1. Oak has *pores*, too, but they are described as being in *rings*, not *diffuse* as in zebrawood. You had not troubled to notice this point before, so you turn back to your veneer. There you find the pores evenly spread, not confined to narrow bands near the start of each annual ring. Oak then is out, and zebrawood remains.

Further, the two-coloured appearance of oak is due to a peculiarity of the rays, whereas that of zebrawood follows the annual rings. A broad and bold pattern denotes zebrawood.

To make sure, turn now to the actual specimens at the front of the book. If you have made the right deductions, you will find your veneer matches your zebrawood specimen closely. Oak and ponderosa pine show up as obvious misfits, as soon as you set them beside the cocktail cabinet for comparison.

6 Wood Names and Their Application

Pilot names

The names used as the main guides to identity in this book are *pilot names*. They have been carefully selected to suit the majority of timber-users in all the English-speaking countries. They belong to the class of *trade names*, discussed later.

Many names, like 'oak' and 'birch', are old-established English names for European woods, which apply equally, in this context, to very similar timbers native to North America. Others, such as 'bubinga' and 'padouk', are taken from the native languages of African or Asian peoples. Others again, for example 'primavera', were given to timbers found in new tropical territories by Spanish explorers.

Several countries, including the United States, Canada and Great Britain, have established *standard names* for both trees and timbers. Our pilot names agree, as closely as possible, with current standard names in all these countries.

When two names are applied to precisely the same timber they are called 'synonyms'. Major synonyms are included in our descriptions of timber, but only pilot names appear elsewhere in the text.

Scientific names

Every known living tree has been given a scientific name in Latin form by botanists who have identified actual specimens of its flowers and foliage. This is the ultimate key to the identity of any timber, but these 'botanical' names are unsuitable for everyday commercial use.

Each scientific name, if given in full, has three parts. The first word, spelt with a capital letter, refers to the *genus* or group of similar trees. The second word, nowadays always spelt with a small initial letter, gives the particular *species* or sort of tree concerned. Both these words are usually printed in *italics*.

The third part, which is printed in Roman type with a capital letter, but often abbreviated, is the name of the botanist, or 'authority', who gave this particular pair of names (genus and species) to this particular kind of tree. Before this name was accepted, he had to examine specimens, describe them in botanical Latin, and have the result published for other botanists to see. If two or more names appear, this is because two or more botanists have played a part in establishing the definitive scientific name. As examples:

Juglans nigra L. is the black American walnut, as described by the Swedish botanist Linnaeus (abbreviated to L.) from the specimens sent to Europe from the United States.

Juglans regia L. is the botanically allied Circassian walnut, also described by Linnaeus from specimens grown in Europe or Asia Minor.

Both these trees come in the common genus *Juglans*, but points of difference appear in Linnaeus's descriptions of their flowers and foliage, and corresponding differences occur in their timbers – as our specimens show.

Sometimes, especially with rare tropical timbers, only the genus is definitely known. The expression 'sp.', set in Roman type, is then used after the generic name, set in Latin. It means 'species uncertain'. The alternative 'spp.', spelt with two 'p's', implies that more than one species is probably involved.

Scientific names always indicate botanical relationship, but pilot names, and other names used in trade, do not always do so. For example, Australian walnut is so close in appearance to the 'true' walnuts that it is sold commercially under that name. But its scientific name of *Endiandra palmerstonii* C. T. White et Francis shows that, botanically, it is quite distinct from *Juglans*.

Note. The scientific name at the head of each Description is that of the actual species represented by the wood specimen in the folder. Other species that are marketed and used under the same pilot name are mentioned in the following text.

Trade names

In timber trade practice, scientific names are used only as an ultimate reference to the identity of a particular sample of timber. One reason for this is that they are cumbersome, hard to pronounce and difficult to spell or remember. Traders like to approach their customers with words familiar to all.

A further reason is that scientific names may be either too *wide*, or too *narrow*, to describe accurately a 'parcel' of timber that a merchant wishes to sell, or a customer wishes to buy. Examples, taken from this book, follow.

(*a*) *Scientific name too wide.* The scientific name *Acer saccharum* Marsh applies to two timbers included here, because botanically they are identical. One carries the pilot name of 'maple', and is the usual, unselected wood of the sugar maple tree, used in solid form for large familiar objects such as furniture, flooring, turned bowls and cotton-reels.

The other timber carries the pilot name of 'Bird's-eye maple', and is specially selected because it holds a remarkably attractive figure due to the presence of buried buds. It is used, in practice, only for decorative veneer and is sold at prices many times higher than those for ordinary maple.

Whatever the botanists may say, the two woods are not interchangeable in commerce. A merchant who sold Bird's-eye maple at the prices prevailing for maple would soon go bankrupt. If he supplied ordinary maple to a customer who demanded Bird's-eye maple, he would have it rejected.

(*b*) *Scientific name too narrow.* Birch is a familiar example of a wood with such constant characters that the timber trade usually lumps several species together. Our specimen is cut from Canadian yellow birch, *Betula alleghaniensis* Britt., found from Ontario and Nova Scotia to the Alleghenies. Two other species grow in North America: the sweet birch, *B. lenta* L., of New England, and the paper birch, *B. papyrifera* Marsh, found right across the continent. In Europe there are two species also: the silver birch, *B. pendula* Roth., and the hairy birch, *B. pubescens* Ehrh. Botanists rightly distinguish these by features of leaf, twig, flower, fruit and bark – hence the five specific names.

But for most of the purposes that birch is required to serve, such as the framing of furniture, cheap tool-handles and ordinary plywood, the wood of all these trees is interchangeable.

Exacting tests show little difference in wood qualities, or in the appearance of their timber under the microscope. So, for most trade purposes, a convenient common name covers all the birches; only when an exceptional need arises is it necessary to define the source more closely.

Names in French, German, Italian and Spanish

A feature of this book is the inclusion of corresponding *pilot names* in practical everyday use in four leading European languages. These appear at the head of the tree descriptions in Part III, and are given in the following form: F. indicates French; G. German; I. Italian; and S. Spanish. If the common name for a growing tree differs from that of its timber, the tree name (or names) follows after a comma. For example, the entry for Douglas fir in French is 'F. Pin d'Oregon, Douglasie'. A few names for tropical timbers, such as 'iroko', are common to all European languages, so no equivalents are possible or necessary.

Certain tropical timbers are imported from several countries where the people speak different languages. Agba, for example, is called by that name in Nigeria, but is known as 'ntola' in the Congo. This explains the origin of its French pilot name, tola blanc, or white tola.

Botanical Family

All the genera of living trees are grouped into botanical families that have broadly similar characters. This is done largely by flower structure, and is too technical for discussion here. Timber structure, though not used for primary classification, follows the same family patterns.

The botanical family name is often of value for further study of related trees and timbers. In its Latinized scientific form – for example, Aceraceae for the maple family – it has international acceptance in scientific circles. It also helps people to visualize the actual trees. Those of the rose family or Rosaceae have flowers resembling wild roses, and their timbers are often sweetly scented. Those of the sweet-pea family or Leguminoseae have flowers like sweet peas, and their woods are often strongly coloured in unusual hues. For these reasons, in the tree descriptions in Part III, the botanical family for the forty timbers has been given in both its English and its Latinized form. It follows the list of European names given at the head of each entry.

The concluding entries before each timber's description are the key number, for ease of reference to Key 1 (which gives the fullest list of identification features), and the source. The source shows the country or countries that yield commercial supplies of the named timber. Where two or more countries are listed, the first name distinguishes the country of origin of the actual specimen included, in so far as it can be ascertained. For timbers with a limited natural range, such as Andaman padouk from the Andaman Islands, it is possible to show the source precisely. But for those trees that grow in a number of countries, for example those of Central America, only a broad indication can be given.

PART III: DESCRIPTION OF FORTY TREES AND THEIR TIMBERS

1 Afrormosia *Afrormosia elata*

F. Kokrodua   G. Kokrodua   I. Afrormosia   S. Olé
FAMILY: Sweet pea (Leguminoseae)   KEY: Eı   SOURCE: West Africa

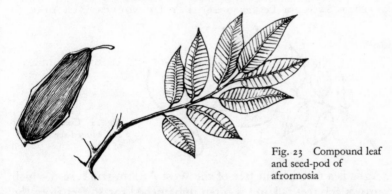

Fig. 23   Compound leaf and seed-pod of afrormosia

This lovely timber brings the African sunlight into northern homes, for its warm crimson-brown colour is lightened by bands of a bright golden-brown shade. Darker bands of dry-season wood make the annual rings distinct. Pores, scattered evenly in small groups through the wood, make the surface lively with a rippling sheen. Afrormosia is hard, heavy and strong, and serves well in heavy building work, as flooring or as ships' decks. It is also widely employed as decorative veneer over furniture, shop-fittings and high-class joinery.

Afrormosia is obtained from a tall forest tree that thrives on open savannas across Africa, from Ghana, Dahomey and Nigeria on the west coast to Chad and the southern Sudan farther east. Local names include 'kokrodua' and 'bonsamdua', while it is also called 'red-bark' or 'devil's tree' because its bark is blotched with red or orange. Its hairy twigs carry large compound leaves, up to a foot in length, made up of seven to nine leaflets. A curious distinctive feature is that the leaflets vary in shape from small broad ovals near the base to large slender oblongs at the tip. Numerous greenish-white flowers, each shaped like a little sweet

89

pea, open on short stalks in open clusters. Each blossom ripens into a large flat pod, about four inches long by two inches across, which is oblong and broader towards its tip; this holds from one to three hard seeds.

The melodious name of 'afrormosia' is a botanist's invention that means 'elm-like tree of Africa'.

2 Agba *Gossweilerodendron balsamiferum*

F. Tola blanc   G. Agba   I. Agba   S. Tola blanca
FAMILY: Sweet pea (Leguminoseae)   KEY: Di   SOURCE: West Africa

Fig. 24   Compound leaf and winged seed of agba

Agba is a magnificent tree of the West African rain forest, which grows 200 feet tall and is often unbranched for 80 feet from the ground. Its base is round, without marked buttresses, and often reaches 20 feet in girth. The leaves are compound, about nine inches long, with about eight pairs of oval, pointed leaflets. The flowers, individually very small, open along slender, branched stalks. The seeds are large, about the size of a pea, with a single papery wing two inches long. Agba is a leading timber tree in the coastal forests from Liberia eastwards through the Ivory Coast, Ghana, Dahomey and Nigeria to Cameroun. Local names include 'noboron' and 'achi', but when exported it is sometimes called 'white tola', 'Nigerian cedar' or 'pink mahogany'.

Agba is actually reddish brown in colour, having a distinct brick-red or terracotta-pink tinge, without any definition of sapwood. Its annual rings are marked by clear, slightly darker bands of dry-season wood, which gives a faint grain. The rays are obscure. Coarse, evenly scattered pores result in fine vessel lines showing on all lengthwise surfaces. Agba has a bright, clean, compact appearance that recalls mahogany, and its working qualities are equally good. It is slightly resinous, and when newly cut has a characteristic peppery smell. Its scientific name implies:

'the tree named in honour of [the botanist] Gossweiler, which bears balsam'.

Agba is light in weight and rather soft. Being easily worked by hand or powered tools, and readily available in large sizes, it is exported on a considerable scale to Europe and North America. Its main uses are for furniture and high-class joinery, including stairways and window-frames.

3 Antiaris *Antiaris africana*

F. Chen Chen, Ako   G. Bonkonko, Kirûndû
I. Chenchen, Ako   S. Chenchen, Ako
FAMILY: Mulberry (Moraceae)   KEY: B1   SOURCE: Tropical Africa

Fig. 25   Simple leaves and berries of antiaris

Antiaris grows as a large, tall tree in the rain forests of West Africa, including Ghana and Nigeria. It often exceeds 100 feet in height, with a girth of 13 feet, measured above the great buttresses that spring out near its base. Since it grows evenly all round the year, there are no clear annual rings in its wood; the pores, which show as coarse vessel lines on longitudinal surfaces, are evenly diffused all through. The general colour is a uniform pale yellow without distinction of heartwood and sapwood. Antiaris timber is exceptionally light, weighing only 27 lb. to the cubic foot, and also remarkably soft. When freshly worked it has a distinct, unpleasant odour.

At first sight this timber would appear to have few merits for export, but as it is available cheaply in large sizes much is shipped to Europe for making plywood. It is also widely employed as a general-purpose hardwood, for such jobs as framing furniture and joinery. In Africa its light weight, softness, and availability in large logs make it a favourite timber for the carving of dug-out canoes.

91

An evergreen, antiaris bears simple oblong leaves set alternately on stout twigs. Its crown branches freely to form a broad flattish canopy. About December it opens clusters of small greenish flowers, which are followed by red fleshy berries, each holding a single seed. The bark is white in colour, smooth except for warty outgrowths, and holds a pale latex that resembles that of the Para rubber tree; unfortunately it has no commercial value. The inner bark or bast is tough and fibrous and was formerly used by the West Africans to make bark cloth. Hence antiaris is also called the 'bark cloth tree'. Toxic substances are found in the bark of this tree, and in the related *Antiaris toxicaria*; and another name, 'upas', signifies 'poisonous'.

4 Ash *Fraxinus excelsior*

F. Frêne   G. Esche   I. Frassino   S. Fresno
FAMILY: Olive (Oleaceac)   KEY: A1
SOURCE: England, Europe, USA, Eastern Canada, Japan

Fig. 26 Compound leaf and seeds of American ash

Even in prehistoric times ash had been recognized as an exceptional tree. It was given an honoured place in the mythology of the Norsemen as *Yggdrasil*, the mighty tree that supported the heavens; below ground its roots went down to hell. Ash is easily recognized by its very large pinnately compound leaf, often nine inches long and bearing, on average, nine pairs of leaflets with a solitary leaflet at the tip. It attracts notice in spring because it is among the last of all trees to open its leaves, often waiting until late May. Its winter buds are also distinctive, being hard, black and set in pairs, on opposite sides of the twigs, with one terminal bud at each twig-tip. The greenish flowers open in feathery clusters just before the leaves appear, and are followed in autumn by bunches of curious winged seeds – the ash 'keys'. These are so called because the shape of a single one-winged seed resembles the metal key used to open a medieval lock.

Ash trees have remarkably open crowns that cast little shade and give little shelter. This is associated with their opposite buds, set in pairs far apart on long twigs and branches. The bark on thin twigs is ash-grey; on older branches and the main trunk it develops a shallow meshwork of ribs and furrows. Records for height are 120 feet for American white ash and 148 feet for common ash in England. The record girth of 19 feet reached in England is, however, exceeded by a tree 22 feet round at Glenn Mills in Pennsylvania. Nowhere is ash considered a really long-lived tree; 200 years is probably its maximum.

Wherever it is found, ash needs really fertile soil to support its fast rate of growth; and for this reason large pure woods of ash are rarely found. It spreads readily by means of winged seeds, which rest for eighteen months in the soil before sprouting. Then they send up oblong seed-leaves, followed by simple leaves, then leaves with only three leaflets. True compound leaves do not appear till the second spring. Ash trees thrive only in full sunlight, but cast little shade themselves.

Several species of ash trees grow in different northern countries. America has three common species, namely black ash, *Fraxinus nigra*, found in the north-eastern States and southern Canada, the green ash, *F. pennsylvanica*, and the white ash, *F. americana*. The two latter kinds grow in most eastern and mid-western States. Europe has its common ash, *F. excelsior*, and Japan the Japanese ash, *F. mandschurica*. Their timbers are so alike that they are used for the same purposes. Ash wood is readily distinguished from that of most other trees. Only American hickory resembles it closely, but can be told apart from a peculiar reddish hue that pervades its timber.

Because of the rapid outburst of large leaves in late spring, the ash tree suddenly needs many large pores or vessels to carry a large current of sap up its stem. So each spring its cambium produces a ring of very large open pores. These stand out at once on every cross-section, appear as conspicuous lines on every radial cut, and show as broad irregular bands on all slash-grain or tangential surfaces. If a wood does not show this feature, it cannot be ash; ash is always markedly ring-porous.

During the summer that follows the fast spring growth, ash cambium lays down very strong, hard, dense summerwood. This makes the wood as a whole heavy and very tough. The faster the tree is growing, the harder and heavier this summerwood becomes, and the greater its value for exacting uses. There are no other remarkable features. The general colour is whitish, without colour distinction of sapwood and heartwood. Despite its strength, ash wood rapidly decays if exposed to damp out of doors.

The toughness of ash ensured its use from earliest times as the best handle for all tools that involve impact, and for most weapons. It also features in sports equipment, for the same reasons. Tools for which an ash handle serves well include hammers, axes, mallets, pickaxes, spades, shovels, garden-forks, sledge-hammers and ice-axes. Weapons of war include spears, pikes, battle-axes, lances, arrows, cross-bow bolts and even the bow itself, though that was usually made of yew. Sports equipment includes hockey-sticks, baseball-bats, oars, parallel bars for gymnasiums, javelins, dumb-bells, tennis-rackets, skis and runners for toboggans and sledges. A valuable quality of ash for all uses that require the grip of the human hand is its smooth surface which seldom splinters.

The same properties of strength and resistance to impact ensure the use of ash in exacting construction. It makes reliable ladder-rungs, and is chosen for the felloes of wooden wheels – that is, the curved pieces that make up the rim and take the shocks from the road. Cart-shafts for horse-drawn vehicles of all kinds are also made of ash, for they too may suffer strain and shock. The framing of carriages, vans and automobiles was formerly of ash, though metals have replaced it.

Ash is a popular furniture timber, because of its supple strength, clean white appearance, and varied though subdued surface figure. It features in much old colonial furniture, sometimes in slats for ladder-back chairs, sometimes as the hoop at the back of the traditional Windsor chair.

To retain the strength of each piece of ash to the full, it is frequently hand-cleft, not sawn. A selected log is split into segments, using an axe and wedges. Each segment is then shaped with hand tools, or turned on a lathe, to its final form. This method ensures strength, because the cleaving tool must follow the wood fibres; it cannot break them as a saw would do, by cutting across. Cleaving also ensures stability of outline, because in cleft material shrinkage is confined within a narrow segment, with the minimum of change in shape.

Ash can readily be bent into curved outlines, without breaking or losing strength. It is first steamed, then bent round a 'former' and held in place by a clamp, then left to dry. Once dry, it sets in its new shape and does not spring straight again. Curved members of furniture, felloes of wheels and ends of hockey-sticks are common examples of bent ash, and so is the curved frame of a snow-shoe.

The word 'ash' is derived from Anglo-Saxon *aesc* and Old Norse *askr*, and is a frequent element in place-names both in Britain and America. Askrigg – 'the ridge of ash trees' – in Yorkshire is a typical example.

5 Aspen Poplar *Populus tremula*

F. Peuplier tremble, Tremble   G. Espe, Pappel, Zitterpappel
I. Pioppo tremula   S. Alamo temblon
FAMILY: Willow (Salicaceae)   KEY: A2
SOURCE: Scandinavia, Northern Europe and Asia, Canada, Northern USA

Fig. 27   Aspen poplar leaves;
note slender leaf-stalk

The aspen poplar tree is easily known by the incessant trembling motion of its leaves, which quiver in the slightest breeze. The mechanism that allows this is a long, flattened leaf-stalk that gives no resistance to a twisting motion. It probably helps the tree by making it easier for it to gain carbon dioxide from the air, through freer passage of currents over the leaf surface. Poplars transpire water faster than all other trees, through open leaf-pores, and 'gas exchange' is easiest when leaf faces are at right angles to wind direction. Aspen is one of the world's hardiest trees, growing in the tundras of all the Arctic regions. During its growing season it needs ample moisture, and this it gains from the thawing of ice locked in the soil. Farther south it is always a streamside or lake-side tree, or grows in bogs. Suckers readily spring up from its roots, and it forms thickets rather than groves of trees.

Aspen belongs to the great poplar genus, *Populus*, found in all the cool zones of North America, Europe and Asia. A constant feature of all poplars is an irregular pattern of branches, buds, and even the veins within the leaf blades. All poplar trees are either male or female, never both. Male poplars open long 'lamb's tail' catkins before their leaves appear; these dull red catkins scatter golden pollen and then fall. Female poplars bear thinner catkins made up of many separate flowers, scattered along a thin stalk like beads on a string. After fertilization in early spring these grow rapidly into green seed-pods. In June these split and release scores of tiny black seeds, each a black grain tipped with a fuzz of white hairs. Those seeds that alight on bare damp earth or mud within a few days of release sprout and strike root; the rest perish. Hence aspen finds it easiest to start life along streamsides.

Most poplars – though not the aspen – take root easily from cuttings. Aspen is increased by seed or by root suckers. Tree-breeders have raised many hybrid strains by crossing male poplars of one kind with females of another. Once a new race is proven it can be multiplied quickly by cuttings or sucker shoots. The giant Swedish hybrid aspen was produced by crossing the common American aspen, *P. tremuloides*, with the European aspen, *P. tremula*, and is now widely planted because it grows far faster than either parent. Another American species, *P. grandidentata*, grows in swamps and streamsides in the north-eastern United States and southern Canada. It is called 'bigtooth aspen' because of the large 'teeth' on its leaves.

All aspen leaves are circular, with a toothed or wavy edge. The bark is smooth and pale grey. Aspen grows fast but rarely forms a large tree, 80 feet being the greatest height.

The wood of all poplars, including aspen, is very distinct from that of other trees. It is white or very pale yellow, without obvious distinction between heartwood and sapwood, or between spring-wood and summerwood. Grain is scarcely seen, and rays are faint. The characteristic feature consists of very large pores, spread evenly through the wood, giving a rough surface – that cannot be planed smooth – on all longitudinal faces. Seasoned poplar is very light, soft and fibrous. It is aptly described as 'woolly'.

The open-pored nature of poplar wood is linked to its rapid growth and high demand for water during a brief, intense growing season. When freshly felled, even in winter, poplar is sodden with sap. It will scarcely float and cannot be made to burn. After seasoning, it becomes one of our lightest timbers, but is a poor firewood because it blazes away too fast.

The main use of poplar is for matchsticks, and aspen is considered best of all kinds. Matchsticks are made by the rotary peeling of large logs into sheets of veneer. These sheets are then divided into thousands of short, square-sided sticks by automatic machines. The sticks are then set on end and dipped in the chemical that makes the match-head, next dried and then fed into matchboxes. The merits of poplar are that it is light, cheap and tough, so it does not snap or splinter when the match is struck. It is also porous, and readily holds the paraffin wax added as fuel. Finally, its ash does not drop or smoulder.

A secondary, but important, use of poplar veneer is as match-boxes and the very light, cheap, 'chip' baskets made by inter-weaving veneers. These are used by the thousand for marketing soft fruit and vegetables. In solid form, poplar is used for cart bottoms because it will not splinter under the impact of stones, and for brake blocks because it does not wear quickly or burn away under friction against iron tyres. Some is used, too, in light joinery.

'Aspen' is an English name related to Norse *osp*. 'Poplar' comes from Latin *populus* and, more remotely, Greek *papaillo*, meaning 'to shake or tremble'. American folk-names are 'trembling aspen', 'quivering aspen', 'shaking aspen' and 'popple'. Scottish and Irish Gaels call it *cran critheach*, the 'shaking tree'. A French Canadian name is *langues des femmes* and a Welsh one *coed tafod merched*. Both mean 'tree of the women's tongues' from its incessant motion.

6 Avodiré *Turraeanthus africanus*

FAMILY: Neem-tree (Meliaceae)   KEY: A3   SOURCE: West Africa

Fig. 28 Avodiré: compound leaf and seed-pods

Avodiré is a beautiful pale wood that belongs to the same botanical family as the mahoganies and shows similar working properties. Nearly white when freshly cut, it darkens on exposure to a pale golden yellow or biscuit colour. Reasonably hard, light yet strong, it shows a fine satiny surface lustre. Its faint, irregular grain gives it, on many surfaces, a fine moiré figure. It is sometimes called 'African satinwood'; local names are 'apaya' and 'lusamba'.

Like many trees of the West African rain forest belt, avodiré has no clear annual rings. Its pores are diffuse and very fine, a feature that distinguishes it from the rather similar, though more yellow, antiaris. Avodiré is a moderate-sized tree, averaging 80 feet tall by 5 feet in girth when mature. Its evergreen leaves are large, up to two feet long, and compound, with about eight pairs of pointed leaflets. Flowers are borne all the year round, being most plentiful in spring and autumn; they are creamy yellow in colour and are grouped in clusters. Each blossom ripens, over a period of six months, an orange-yellow fig-shaped pod, about one inch across. This is fragrant and has a soft fleshy pulp within, holding four or five seeds. The bole of the tree is often crooked or misshapen; it bears a white bark that is reputed to be poisonous.

7 Beech *Fagus sylvatica*

F. Hêtre   G. Buche   I. Faggio   S. Haya
FAMILY: Beech (Fagaceae)   KEY: D2
SOURCE: England, Europe, USA, Japan

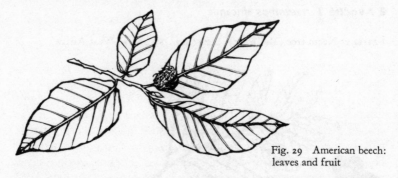

Fig. 29   American beech: leaves and fruit

Beech trees are easily known by several clear characters. Their bark is remarkably smooth – no matter how old the tree – and metallic grey in colour. Winter buds are long, slender, clad in brown papery scales and always set singly, usually at bends on the twigs. The leaf is a perfect oval with an almost smooth edge – a simple design rare in nature. The flowers, greenish yellow, are grouped in catkins of male or female sex, and the latter ripen to small triangular brown nuts, set in green spiny husks. Beechwoods are remarkable for their dense shade in the summer months. Few flowering plants, grasses, ferns or even mosses can grow in a beech tree's shadow, and the forest floor is carpeted with drifts of brown faded leaves that slowly decay to form a rich mould. Beech usually forms pure woods, for seedlings of other trees cannot start life in its shade. It thrives on moderately fertile soils and tolerates both chalk and limestone.

The cambium of the growing beech tree produces wood of a remarkably even texture throughout the spring and summer. The darker summerwood of each annual ring is visible but never conspicuous. The pores are very fine and are spread evenly through the wood – which is typically diffuse-porous. The general ground colour is pinkish or reddish brown, without clear distinction between heartwood and sapwood.

A small but constant and conspicuous feature makes every piece of beech wood 'two-coloured', and enables you to name it easily. This is the ray figure, for the rays have a characteristic chocolate-brown shade and are scattered evenly all through the timber. They show as clear tiny specks on the slash-grain surface, and as little flecks or flakes on radial cuts. If a specimen does not show these little brown marks, it cannot be beech.

Beech wood is hard, strong and heavy, but it lacks the toughness of ash and cannot be used for long handles subject to shock, since it proves too brittle. It has no natural durability in contact with the ground and is seldom used out of doors.

Indoors it proves a most serviceable and versatile wood for furniture, short tool-handles, flooring and the hundreds of jobs that call for 'a piece of wood'. Because of its even growth and lack of both large pores and large rays, it is easy to work in any direction – either with, across, or at any angle to, the grain. Few common timbers have this property, and it makes beech a particularly suitable wood for modern mass-produced goods that can be cheaply made on high-speed machines. Lack of pronounced grain also gives a smooth even surface that resists wear well.

Beech is easily bent to new shapes after steam treatment, and many thousands of bentwood chairs are made annually by curving straight lengths of beech into half-round backs or curved legs. It is very suitable for laminating, and when rotary-peeled it yields large sheets of an attractive and serviceable veneer, widely used in beech-faced plywood. Branchwood and waste make first-rate firewood. Supplies are good and large wood-working factories in America, Britain, Europe and Japan depend on beech as their key raw material.

Traditional uses for beech include wooden plates, bowls and platters, turned by hand on simple lathes worked by foot power. Turners also made chair-legs, wooden spoons and tool-handles. In the Chiltern Hills of Buckinghamshire, beech 'bodgers', as they were called, would fell a tree, saw and cleave it into segments, and turn these into chair legs without ever leaving the woods. They used a hand-made pole lathe, unchanged from the remote past, to give a rotary motion to each piece of beech that they shaped with their chisels. The last bodger survived until 1950.

Beech can be found in every craftsman's tool-kit, though plastics nowadays compete. Handles of chisels, screwdrivers and mallets that must take considerable but not extreme shock and strain are usually made from beech. Its smooth even surface is kind to the hands. Mallet-heads, and many similar small wooden tools, are also shaped from beech. Every wooden carpenter's plane is made from beech, partly because of its strength and smoothness, but mainly because, once well seasoned, it remains very stable, enabling the worker to plane a true and accurate surface.

Beech plays a large part in everyday furniture, such as school desks, kitchen tables and workshop benches, because it is plentiful and fairly cheap, hard, smooth and stable, with a bright clean appearance. A great deal is used in leisure furniture, but often out of sight in framing and the support of decorative veneers. It is used in footwear for sabots, sandals and heels. Oddments of wood, such as the backs of cheap brushes, or darning 'mushrooms' for the housewife, are made of beech in great numbers, and so are some very small objects, such as cocktail sticks. In densified form, impregnated with plastic resin, beech is used for the handles of kitchen knives, etc., and as plywood for trays; it then appears chocolate-brown, smooth, heavy and hard.

Lack of obvious figure limits the employment of beech as an ornamental wood, but flamy veneers can be very attractive. It makes excellent floor-blocks.

American beech, *Fagus grandifolia*, has a natural range from Nova Scotia and Ontario southwards through all the eastern States to Louisiana and parts of Texas; but to the west of the Mississippi it is rarely found except as a planted shade tree. In the wilds it forms dense woods and grows 120 feet tall. Europe has a single species, *F. sylvatica*, which grows usually on the foothills of the higher mountain ranges. Beechwoods form a definite zone on the lower slopes of the Alps, the Pyrenees, the Apennines, the Sudeten Mountains and the Carpathians, and timber from this tree is a major resource in a dozen countries. France, Spain, Italy, Germany, Switzerland, Czechoslovakia, Austria, Yugoslavia, Romania, Poland, Belgium and Denmark all have major wood-using factories based on beech. In England it is native to the southern and Midland counties only, forming beautiful beech-woods on the chalk downs, the Cotswolds and the Chiltern Hills. These support a big furniture-making trade round High Wycombe in Buckinghamshire. Beech also grows in Turkey and on the Caucasus Range of south-east Russia, while Japan has its distinct species, *F. crenata*.

Beech has been widely planted as an ornamental tree, parti-cularly on big country estates in Scotland and Ireland. The copper beech, *F. sylvatica* 'Purpurea', which arose as a natural sport in Germany about 1780, has the green colour of its leaves masked by coppery-red pigments. Though usually grafted, it grows true from seed; its wood cannot be told apart from that of common beech.

8 Birch *Betula alleghanensis*

F. Bouleau   G. Birke   I. Betulla   S. Abedul
FAMILY: Birch (Betulaceae)   KEY: B2
SOURCE: Eastern USA, Canada, Northern Europe, Northern Asia

Fig. 30   Birch foliage
and fruiting catkin

Birch is known to all as a lovely white-barked tree with slender branches and fine twigs that carry myriads of dangling bright green diamond-shaped leaves. The world's hardiest tree, it grows right round the polar regions, with dwarf races struggling for life in the Arctic tundra, on the perma-frost soils that are frozen over most of the year, and that never – even in summer – thaw out for more than a few feet below the surface. Farther south, birch is a tree of the high mountains and the waste lands, quick to colonize any abandoned farmland or newly exposed streambank or avalanche track by means of its light wind-borne seeds. Foresters rarely plant birch, for there is always more than enough springing up of its own accord.

The young slender twigs of birch are purplish brown in colour, and bear a white waxy bloom in spring. The distinctive white bark is produced later, by the bark cambium, on all substantial stems. At the base of old trees it eventually becomes fissured and turns black. The white colour of birch bark is probably a protective device to keep the tree-trunk and the sap within it cool. Though we think of the northland as cold, its brief summer is intense. The sun's rays strike at a low angle, from each direction of the compass in turn, for twenty-four hours a day. In the open northern woods a birch trunk receives more radiation, during its active growth spell, than a teak trunk in the dense shady tropical jungles.

Birch bark is full of natural waxes that make it waterproof. It is also tough and durable, and persists in the soil long after the wood within it has rotted away. The Lapps use pieces of birch bark as roofing shingles for their huts, or as plates, and shape them into circular boxes. The Swiss use birch bark to make their huge, but remarkably light, musical alphorns. Similar crafts were developed by the North American Indians, who have available the tough bark of the paper birch. After felling a selected straight tree, they peel off its bark in large sheets, and fasten it over a light wooden frame to make their amazingly light birch-bark canoes. Originally, the sheets of bark were sewn with spruce roots and caulked with resin from a silver fir. Wigwams were also built of birch bark, by securing plates of it to light poles.

Male catkins of birch are conspicuous in spring, when they hang down as yellowish-brown 'lamb's tails' among the opening leaves. At this time the female catkins, which are much smaller, green and erect, are seldom noticed. By autumn, when the leaves are fading to gold, ripening fruit catkins attract the eye as rather plump 'lamb's tails'. These soon break up completely, each catkin releasing hundreds of bracts that fall to the ground, and hundreds of tiny winged seeds that are carried away on the lightest breeze. Millions of birch seedlings sprout each spring, but the tree is a short-lived one. It rarely exceeds 80 years of age, 80 feet in height or a girth of 4 feet. At all times it must have full sunlight. Birch can tolerate poor soil, but never shade.

The timber of birch is the most featureless of all northern hardwoods, and this in itself aids identification. Rings, grain and pores are all alike obscure, and the colour is an even pale brownish yellow without distinction of heartwood and sapwood. As usually sawn and seen, birch has a dull lustreless surface, due to the even scatter of fine pores or vessels all through its substance. It is definitely hard, dense and moderately heavy. Out of doors it has no natural durability.

The bulk of the world's birch is still used for firewood. The alternatives in the northern or mountainous regions where it thrives are conifers or softwood trees, and birch is found to give a far more lasting and intense heat, both for cooking and in house-warming stoves. The stack of round and cleft birch logs remains a familiar sight beside the homesteads of Norwegian and Swiss farmers, as it used to be in North America before the coming of oil, coal and electricity. Regrowth on the farm or community woodlot ensures a perpetual fuel wood supply. Birch is also a first-class charcoal wood, and was for long a key raw material in the wood-based iron-smelting industries of North America and northern Europe. Swedish ironmasters used birch charcoal for making high-grade steels right down to 1960.

Northern peoples use birch for many everyday jobs, partly because of its physical properties but largely because they had, in the past, no alternative hardwood. In Finland, for example, plain but cheap and serviceable furniture is constructed from native birch, and so are trays, bowls, boxes and tool-handles. Birch, though hard, is easy to carve or turn to a smooth finish. One wide-spread use is as broom-heads, which are turned as round cylinders and then sawn into two half-cylinders to take the bristles and the handle. Cheap household brushes are made in the same way.

The trunks of large birch trees yield a valuable supply of veneer for cheap plywood, manufactured on a large scale in America and Scandinavia. In typical three-ply wood the centre layer is made of spruce veneer, set with its grain at right angles to the two outer layers, both of birch. The birch contributes hard firm surfaces and some rigidity. Far more costly decorative veneers are cut from selected stems that show exceptional figure, including flame, bird's-eye and the intricate tracery of Masur birch from northern Scandinavia.

Even birch twigs have their practical uses, besides featuring in school discipline and witches' broomsticks. Handy brooms for sweeping lawns and yards are made by fastening leafless, well-seasoned birch twigs, cut in winter, in tight bundles on straight handles. In Scandinavia much smaller bundles of white, barked twigs are used for whisking eggs and sauces. Vinegar manufacturers employ layers of birch twigs as a medium to support bacteria which turn alcohol into acetic acid. Metal-refiners once used them to clean copper and tin-plate because they turned to charcoal when heated, and this in turn removed harmful oxides and oxygen from the hot or molten metal.

The characteristic birch of Canada and the northern States is the paper birch, *Betula papyrifera*, which has a tremendous range from Labrador and Newfoundland east to Alaska. Sweet birch, *B. lenta*, is frequent in New England, and its range extends southwest down the Appalachians to Alabama. Yellow birch, *B. alleghanensis*, which draws its name from a yellowish tint in its silvery-grey bark, is found in the St Lawrence Basin, in the Lake States, and on the upper, cooler slopes of the Alleghenies and Appalachians.

In Europe and northern Asia there are two birches, the silver birch, *B. pendula*, with smooth purplish twigs studded with little warts, and the hairy birch, *B. pubescens*, which has downy twigs. A very beautiful form with pendulous branches and irregularly lobed, long-pointed leaves, *B. pendula* variety *dalecarlica*, originates in the Swedish province of Dalecarlia and is often cultivated in gardens.

9 Bird's-eye Maple *Acer saccharum*

F. Erable œil d'oiseau   G. Vogelaugenahorn
I. Acero occhiolinato   S. Arce ojo de pájaro
FAMILY: Maple (Aceraceae)   KEY: A4   SOURCE: Eastern Canada, Eastern USA

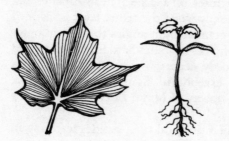

Fig. 31   Bird's-eye maple: leaf and seedling

Bird's-eye maple is a remarkable form of white or sugar maple (described on p. 131), which shows a particularly attractive figure. Most of its physical properties resemble those of ordinary maples, but it shows two colours, the whitish ground being broken at irregular intervals by brownish dots. These dots are seldom solid. Usually they have a circular rim differently coloured towards the centre, rather like an eye, and this explains the timber's name.

How do these circular dots arise? Basically they are 'bud initials', the starting-points for fresh side branches that may, or may not, have actually grown out from the trunk of the tree. Many broad-leaved trees can produce two sorts of branches. Main branches start from small twigs, and as they get buried when the trunk grows stouter they appear as *knots* when it is eventually sawn up (see p. 34); they are few in number and are set well apart. The second kind of branch is called an 'epicormic' shoot or sprout, and it starts well out in the stem. Sprouts seldom develop unless the main trunk is damaged; they are a kind of second string to enable a tree to repair an injury such as windbreak, or to grow vigorously once more after severe pruning.

It is these secondary branch-bud initials that make the 'bird's-eyes'. Since they lie all round the circumference of the tree-trunk, and point outwards, they only show their characteristic shape if the wood is cut at right angles to the radius of the stem. This is usually done by rotary peeling. Bird's-eye maple is in great demand for its attractive effect, so logs that will yield it are always used for veneer. It is not easy to detect it, until the log is cut, but experts usually suspect its presence from surface irregularities or the actual appearance of small branchlets. Similar figure is found, though rarely, in other broad-leaved timbers, such as birch.

10 Bubinga *Guibortia demeusei*

FAMILY: Sweet pea (Leguminoseae)   KEY: C3   SOURCE: West Africa

Fig. 32   A bubinga leaf consists of two leaflets; right, a seed-pod

Bubinga is a fascinating wood that originates in Equatorial West Africa, in Nigeria, Cameroun, Gabon and the Lower Congo. In the northern parts of its range it is called 'bubinga' and in the southern parts 'kevazingo', both being native names for the tree that yields it. In timber-trade practice, however, the name of kevazingo is reserved for rotary-cut veneers which display its exotic colouring to best advantage, and wood prepared in any other way is called bubinga.

Bubinga has a purplish-brown ground colour, bordering in some specimens on deep crimson. Deeper tints run across it, either as irregular bands or as mottled or marbled variegations. This figuring is independent of the distinct annual rings, shown by darker summerwood. Fine pores are diffused all through the wood, and usually hold a reddish gum; a resin known as 'gum copal' can be obtained from closely related trees of the genus *Copaifera*.

Bubinga grows as a slender tree scattered through open savanna woodland amongst many other species. It rarely exceeds 70 feet in height or 9 feet in girth. Its white sapwood is of no decorative value, and logs with enough coloured heartwood for economic harvesting are scarce and valuable. The bark is grey and rather thick. Leaves are remarkable in consisting of only two leaflets each, set side by side and having oval, curved and pointed outlines. Small flowers, resembling sweet peas in structure, are scattered along a slender flower-spike, and develop into a string of seed-pods. Each separate pod is papery, of a curious triangular-oblong shape, and holds several small hard seeds.

Bubinga, and its rotary-peeled form kevazingo, are used as decorative veneer for panelling, shop-fittings and furniture, where an unusually striking effect is desired. Solid wood is used for pianos, cabinet-work and small ornamental articles such as brush-backs, knife-handles or hand-carved woodware.

11 Cedar of Lebanon *Cedrus libani*

F. Cèdre du Liban   G. Libanon-Zeder   I. Cedro del Libano
S. Cedro de Libano
FAMILY: Pine (Pinaceae)   KEY: E2
SOURCE: England (introduced)
Near East, USA (introduced)

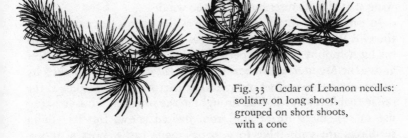

Fig. 33   Cedar of Lebanon needles:
solitary on long shoot,
grouped on short shoots,
with a cone

Cedar of Lebanon is well known to everyone who has read his Bible, since it was in fact the wood used by Solomon to build his temple at Jerusalem, about the year 1000 BC. The Hebrew name for both tree and timber is *erez*, with the plural form of *arazim*. We draw our name of 'cedar' from Greek *kedros*, and Latin *cedrus*, words originally implying a fragrant juniper bush, but later extended to sweet-smelling woods of many kinds.

The true cedars, as trees of the botanical genus *Cedrus* are now called, are remarkable conifers that survive only on high mountains in the sub-tropics. The deodar or Indian cedar grows on the Himalayas, the Atlas cedar on the Atlas Mountains of Morocco, the Cyprus cedar on the Troodos Range in Cyprus, and the Lebanon cedar on Mount Lebanon and a few other higher mountains of Lebanon, Syria and Turkey. It cannot flourish in the warmer lowlands, and this explains why Solomon bought his cedar timber from Hiram, King of Tyre, in exchange for wheat and oil. The logs were hauled to the sea and floated for 200 miles down the Mediterranean coast, then carried inland to Jerusalem.

All the true cedars are evergreen, with two arrangements of needles. On their slender, terminal shoots the needles are placed singly, while on the more numerous side shoots they are grouped in rosettes. Each side shoot produces a fresh rosette every year, but it elongates extremely slowly – a shoot one inch long may be twenty years old. Male flowers do not open until August, and then appear as conical catkins that shed golden pollen from numerous anthers, and then fall from the twigs. The female conelets, which are pollinated at this time, are green, erect and oval. They take two years to ripen into curious barrel-shaped woody cones, about four inches long by two inches wide. These cones stand upright on the twigs. They have broad scales which slowly fall away, to release the winged seeds gradually during the year that follows. Seeds are borne two to each scale; each is about the size of a grain of wheat, and has a much larger triangular wing that aids its distribution by the wind.

In prehistoric times cedars were widely spread over much of the Old World, but climatic changes have isolated the four kinds on high mountains with peculiar climates. Lebanon, 6,000 feet above the Mediterranean Sea, is snow-clad in winter but baked by fierce sun in the summer months. The narrow, waxy needles of the cedar enable it to withstand drought in the summer, whilst making use of soil moisture arising from melted snow. But its young seedlings are vulnerable to grazing goats, cattle, horses, sheep, camels and asses, and where the woods are open to wandering livestock they cannot survive. In consequence the cedar groves were steadily decreasing until the present century, when they received protection from the Lebanon Government – partly for their historic interest, but largely as a tourist attraction.

Veterans on Mount Lebanon reach 80 feet in height, with trunks over 40 feet round. Growth is slow at such altitudes, and some may be 2,000 years old – veritable survivors from Biblical times. Lebanon cedar is distinguished from other nearly related kinds by its very flat habit of branching – an outline familiar from Lebanese postage stamps. If you gaze upwards through its ever-green canopy, it looks like a pattern of elegant lace. At one time the groves were owned and venerated by a sect of Maronite Christians, who held an annual religious festival in their midst.

From the days of the Crusades the cedars of Lebanon have been visited by pilgrims and travellers from the West. Many carried cones and seeds back with them but the date of the cedar's successful introduction to Europe is unknown. Probably the Reverend Edward Pococke (1604–91), who was chaplain to the English Turkey Merchants in Aleppo from 1630 to 1635, was the first European gardener to raise it; he planted a tree at Childrey in Berkshire in 1646, which still survives. No trees in England are more than 350 years old, but some have reached great size – 140 feet high at Foxley in Herefordshire and 38 feet round at Cedar Park near Cheshunt in Hertfordshire. Nearly every great park in England now has its cedars, and they are naturally a favourite tree for rectory gardens. Lebanon cedar is equally popular for landscape planting in North America, and indeed throughout the temperate lands of the world.

The bark of Lebanon cedar is dark grey and rather thin; it is broken into squarish plates by shallow fissures. Below it lies the thin zone of whitish sapwood. The heartwood is, overall, a soft, warm brown colour, and shows a distinct division into darker and denser summerwood and paler springwood bands. The timber as a whole is soft and remarkably light. Nevertheless, the heartwood is naturally durable, thanks to the natural oil that pervades it. This has a characteristic fragrance resembling incense, and is distilled from the wood for use in perfumery. Besides oil, cedar wood holds resin, secreted in fine resin canals.

A curious feature that aids identification is the irregular edge of each annual ring. In most trees annual rings form true and regular circles, but in cedars the rings are wavy or rippled at their outer, summerwood edge. This adds to the attractions of the grain.

Cedar wood is easy to work and strong enough to serve a wide range of uses. In the Himalayas, where Indian cedar is plentiful, it is used for house-building, railroad ties, furniture, bridge construction and general carpentry. But the timber of the cedar of Lebanon is too scarce for such everyday jobs, for the only sources are the limited eastern groves and occasional specimen trees that have to be felled in gardens. Large logs, when available, are usually reserved for radially-sliced veneers, which have a pleasing effect due to the regular repetition of the naturally wavy grain.

12 Cedar, Western Red *Thuja plicata*

TIMBER: F. Cèdre rouge   G. Rotzeder   I. Cedro rosso   S. Cedro dulce
TREE: F. Thuja géant, arbre de vie   G. Riesenlebensbaum
I. Arbor vitae   S. Thuya
FAMILY: Pine (Pinaceae)   KEY: D3
SOURCE: British Columbia, Alaska, Western USA, Europe (introduced)

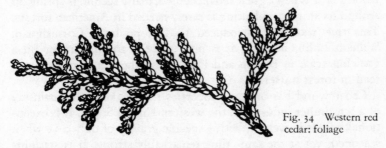

Fig. 34   Western red cedar: foliage

This attractive and important timber gets its name from two features that are very obvious when a log is freshly cut. It exudes a strong aromatic fragrance, recalling the cedars of Lebanon (p. 106), and shows a bright red-brown surface. Eventually, as it seasons and matures, its characteristic scent is lost, and it weathers to a silvery grey.

The tree that yields it grows only in a limited zone around the Rocky Mountains, and for this reason it is called the *western* red cedar. It is found from Sumner Strait in Alaska, south through British Columbia, Washington and Oregon to the north-west corner of California, and inland as far east as Montana and Idaho. Other well-established names are 'stinking cedar', 'shingle-wood', because it makes first-class roofing shingles, and 'canoe cedar', because Indians used it for making dug-out canoes. When grown as an ornamental tree or hedge shrub, it is called the 'giant arborvitae', or 'tree of life', because of its cheerful evergreen foliage.

Western red cedar is a conifer, but its flowers and cones are small and rarely attract attention. Its foliage resembles the fronds of a fern, for the leaves hide both twigs and buds; they are bright green and give out a strong, sharp scent when bruised. Western red cedar and the allied eastern white cedar, which proves hardier in districts with frosty winters, are widely grown as decorative shrubs and hedges in both America and Europe. Indoors, their cut foliage is popular for flower arrangements. It is also widely used for wreaths.

In its native forests the western red cedar forms a magnificent tree with a stout buttressed trunk clad in fibrous reddish-grey bark. Some giants reach 200 feet in height with girths up to 30 feet, and ages of 400 years. They are free from side branches for many feet up. This means that their outer layers are knot-free or 'clear' and particularly valuable for high-class joinery and woodwork. When a great cedar is felled, many seedlings spring up around its stump; replanting is rarely needed in American forests. This tree was first introduced to Europe by a Cornishman, William Lobb, in 1853; it is now planted for its timber on a growing scale in Britain and France and there it is raised from seed in forest nurseries.

Cedar wood has unique properties that make it the mainstay of a big timber industry in the western United States. It is exceptionally light in weight, with a specific gravity of only 0·38 when seasoned, yet at the same time remarkably strong. It is straight-grained and only moderately hard, which means that it can be worked to very precise dimensions, with a good surface finish, by all kinds of hand and machine tools. Finally, it is naturally durable, and resists decay out of doors, under most climates, without any kind of preservative treatment. The British Isles are an exception, since their mild winters favour decay, and an occasional surface dressing of creosote is needed to prolong its service life.

For many exacting purposes western red cedar has no equal. It is used throughout America and exported to Europe on a large scale, commanding a higher price than all competing softwood. It is the best wood for ladder-poles, where its light weight, strength, straight grain and freedom from knots all make for handiness, safety and reliability. It is used for building bungalows, wooden sheds and greenhouses, for it can be machined to exact dimensions that remain stable. This makes it particularly suitable, too, for window-frames and barrel-staves. Shingles of cedar wood, once all cleft by hand but nowadays machine-sawn, are the most popular roofing material right across America. They are lighter and cheaper than tiles and remarkably durable. A new cedar building stands out at first because of its bright orange-brown hue, which later mellows to an attractive silver-grey. On farms cedar is used for posts and poles, because of its durability, and also for fencing and building barns.

The natural durability and smooth working properties of red cedar were early recognized by the Indians of the Pacific coast. They made their great war canoes by hollowing out selected cedar trunks, and it was the only timber used for carving totem-poles. They knew it would endure indefinitely, preserving the brightly coloured features of their tribal ancestors. Modern wood-sculptors still employ it, and it makes excellent trays and tableware, light in weight with a pleasing grain and surface.

The heartwood is a warm red-brown, while the thin sapwood zone is creamy yellow.

The Haida Indians, who lived along the densely forested shores of southern Alaska, gained strange but serviceable raw materials from cedar trees. They treated the thick bark as a textile fibre, and wove it into ropes, baskets, mats, clothes and hats. The tough twigs were used to make stouter ropes for whaling, while arrows were carved from still stouter stems. Lacking steel, they used the tough knots for fish-hooks!

Western red cedar belongs to the genus *Thuja*, a curious name first used by the Greeks for a related tree, nowadays called *Tetraclinis articulata*, or in Spanish *alerce*, which grows in North Africa and yields a fragrant gum called 'sandarac', used to make varnishes. Both belong to the cypress group or Cupressineae of the pine family. Two of the six species of *Thuja* grow in America, while the other three are native to China, Japan, Korea and Formosa.

13 Cherry *Prunus avium*

F. Merisier, cerisier   G. Wildkirsche, Kirsche   I. Ciliegio   S. Cerezo
FAMILY: Rose (Rosaceae)   KEY: E3
SOURCE: Europe, USA, Canada, Japan

Fig. 35   Cherry foliage, fruit and flower

Many kinds of cherries flourish in forests, gardens and orchards, across America, Europe and Asia. They reach their peak of spring beauty in Japan, which has provided a wealth of ornamental flowering varieties for parks and gardens throughout the temperate lands of the world. Both these and the cherries grown for fruit are usually hybrids bred by skilful gardeners and propagated by grafting. Few grow stout enough to yield useful timber, and most cherry wood in commerce comes from a single European species, *Prunus avium*.

In England this wild tree is called the 'mazzard' from an old French name, *merisier*, applied to all kinds of cherry. The Scots call it the 'gean', from an old Italian word, *guina*, which originally meant a cultivated cherry. Its Latin name, *Prunus avium*, means 'the plum tree of the birds', but in practice the cherries are distinguished from the plums because they bear their flowers in clusters, never singly. The birds relish their fruit.

In the woods of northern Europe, including Britain, wild cherry grows singly or in small groups, often associated with beech. Though never very stout – the record is 12 feet round – it may grow tall, sometimes attaining 100 feet in height. It is one of the parents of orchard cherries, and this explains why those trees need constant top pruning to keep them a reasonable size. Cherry trees are known at once by their smooth lustrous red-brown bark, which is broken at frequent intervals by raised patches of cork. These are the breathing pores or lenticels, which cherry needs to get air into its trunk through an otherwise impervious layer. On old trees the bark breaks away in thin horizontal strips.

The winter buds, placed alternately on brown twigs, have many visible scales. The leaves, which open in late April or early May, are long-stalked, simple and oval in outline, with toothed edges and a distinct point. They turn to vivid shades of orange and scarlet in the autumn, and at that time a cherry tree stands out amid darker beeches like a burning bush. If spring is late and cold the white blossom bursts out while the tree is still leafless, decking the whole crown of the tall trees with white petals over a dark tracery of twigs, like a mantle of snowflakes. In a warmer, earlier spring, white blossom harmonizes with the delicate brownish-green of opening leaves.

Cherry flowers open in bunches on long stalks, which spring only from short shoots or spurs along the smaller branches. Each flower has five green sepals, five white petals, a multitude of golden stamens and a one-celled ovary that develops a single fruit. This fruit, the familiar cherry, changes from green through white and red to black as it ripens. By late June it is fully ripe and it then attracts the birds, although its flesh is too thin to attract humans. Thrushes swallow the cherry, digest the seed, and void the hard stone, which sprouts next spring. But seed that falls normally has to lie on the forest floor for eighteen months before it will start to grow.

Cherry wood is readily told apart from other brownish woods by its golden sheen, carrying a hint of green. It brings the sunshine of the forests with it, and also their fragrance, for newly worked timber smells like rose blossom. There is a sweet-smelling gum all through the wood, and this pervades any room where cherry logs are burnt as firewood. The grain is very lively, for scattered pores and tiny rays reflect the light like a myriad of tiny mirrors. Thin dark brown bands of summerwood make the annual rings distinct. Sapwood, if present, has a pinkish hue.

Cherry is a hard though fairly light timber, with good working properties. Solid timber is used for high-class furniture, picture-frames, decorative turned woodware such as fruit-bowls, ornamental boxes, clothes-buttons, furniture knobs, shop-fittings and similar high-grade joinery. Very attractive veneers are made by slicing cherry logs radially. They are used on beds, dressing-tables and similar large pieces of the finest furniture. Cherry deservedly enjoys high prices and a keen demand. Most of the brightest-figured wood is drawn from forest trees; old orchard stems occasionally reach the timber merchant or veneer-cutter, but most are only large enough for firewood.

14 Douglas Fir *Pseudotsuga menziesii*

F. Pin d'Oregon, Douglasie G. Douglas, Douglastanne
I. Pino d'Oregon, Douglas S. Pino d'Oregon, Abeto de Douglas
FAMILY: Pine (Pinaceae) KEY: D4
SOURCE: British Columbia, Alaska, Western U.S.A., Europe (introduced), Australia (introduced), New Zealand (introduced)

Fig. 36 Douglas fir needles and cones, with three-pointed bracts below every scale

Douglas fir is one of the most magnificent trees in the world – indeed it once held the height record. In 1895 loggers felled a tree in the Capilano Valley, a few miles north of Vancouver City in British Columbia, which was scaled – before the Mayor, the Sheriff and other reliable witnesses – at 417 feet. But its equal has never been encountered since, and today the record is held by a Californian redwood, *Sequoia sempervirens*, in Redwood Creek Grove, Humboldt County, at 368 feet. The tallest living Douglas fir recorded in the *National Geographic Magazine* (July 1964) is at Ryderwood, Washington, and attains only 324 feet; but another, felled at Mineral, Washington, scaled 385 feet and once outclassed all the sequoias. The greatest girth recorded, at Clatsop County, Oregon, is 49 feet. Many large Douglas firs have been found, by ring counts, to exceed 1,000 years in age.

Douglas fir is easily known by several pointers. On young stems the bark is smooth and bears prominent resin blisters, which exude sticky, clear-yellow, fragrantly scented resin when broken. On older stems the bark becomes exceptionally thick and rugged, with deep fissures and long irregular ribs between them. Minor corrugations show a curious cinnamon-red shade, distinctive to this tree. The needles are borne singly on the twigs, and if they are pulled away they leave a neat circular scar; this distinguishes Douglas fir from spruces, which carry their needles on woody pegs. The buds are slender, pointed and non-resinous, with brownish scales rather like those seen on beech; this marks them out from most other firs, which bear blunt, resinous buds. Male flowers, opening in June, have the typical conifer design of short-lived clusters of yellow anthers. But the female conelets, which ripen in six months to woody cones, show from the outset the unique Douglas fir symbol. This is a three-pointed bract, rather like a trident, which peeps out from the base of every cone-scale. The cones are egg-shaped and about three inches long by one and a half inches wide. They open in autumn to release winged seeds, two per scale. Young firs spring up very freely in the forests of the West, even though wood mice and birds eat many seeds.

Douglas fir grows as a native tree southwards from British Columbia through all the western States to the Mexican sierras. Its range extends inland to Alberta, Montana, Wyoming, Colorado and New Mexico, following the Rocky Mountains, but the largest exploitable stands are in British Columbia, Oregon and Washington. It has been introduced as a promising timber tree to many countries overseas, and is widely planted in Great Britain and Ireland, France, Germany, Belgium, New Zealand and Australia. The tallest tree in Britain is a Douglas fir, 182 feet tall, at Powis Castle near Welshpool in Wales.

The Latin specific name, *menziesii*, records the Scottish botanist, Archibald Menzies, who sailed with Captain Vancouver's exploring expedition in 1794, and sent the first specimens to European botanists. Seed was sent eastwards later, in 1827, by David Douglas, another Scottish botanist whose name has been linked ever since with this magnificent tree. But timber merchants exported its timber to Europe as 'Oregon pine', because of its pine-like character, and this name is still used by timbermen in many European lands.

The exploitation of the coastal Douglas fir forests, begun in the mid-nineteenth century, is still going on. Modern State and Federal controls ensure the replacements of felled stands, either by supervised natural seeding or by artificial planting. These magnificent forests are a resource of great economic value to Canada and the United States, and will continue to be so in the future. The large size and high value of their timber call for the use of the world's most powerful and costly logging equipment, including enormous sky-lines, loading cranes, trucks and rafts for coastwise floating.

Douglas fir timber, like all softwoods, lacks pores and vessel lines. Like most softwoods it is resinous, with the characteristic smell of turpentine. Resin exudes from any cut on the living tree and – after the turpentine has evaporated – leaves a coating of yellow rosin as a protection against insect or fungal attack. Douglas fir lumber shows a remarkably clean-cut division between the hard, red-brown summerwood bands and the paler, softer, pinkish-yellow springwood. Another feature peculiar to Douglas fir is the even thickness of the summerwood band, relative to the neighbouring springwood of the same annual ring. No matter how fast the tree may grow, it always lays down a high proportion of strong, dense summerwood. This makes it a very satisfactory and dependable timber for heavy constructional work.

Douglas fir is, overall, soft and easy to work, though the marked difference between summerwood and springwood call for special care with cutting tools. It is light yet reasonably strong and firm, but not durable out of doors unless treated with preservatives. The sapwood layer is almost white, but being rather thin is seldom seen on processed lumber.

The good physical properties of Douglas fir, and its availability in large quantities and big sizes, have led to its use for a wide range of purposes, both locally on the West Coast and after export to the eastern States or to Europe. It is applied to every kind of heavy constructional work, in house- and factory-building, railroad and dock engineering, and also in bridging and shipbuilding. It is used for railroad ties, mine props, telegraph poles, transmission poles for electric power, packing cases, indoor joinery, general carpentry and farm fencing. Vats, water-conduits, ladders, barrels, and ships' masts are also made from it. The 'clear' knot-free lumber cut from the base of large old trees is particularly strong and valuable.

Large Douglas fir logs are particularly well suited for the cutting of rotary veneers for making plywood in large sheets, and of high strength, for structural use. Such veneers display the strong grain to advantage, in large surfaces, of contrasting colour and texture, having irregular, often flame-shaped, outlines. Natural colour is so strong that staining is not needed, but textured effects are sometimes added by sand-blasting. Strong yet decorative doors and panelling are made on a large scale from Douglas fir plywood.

Other uses include pulpwood for paper-making, hardboard, chipboard, insulation board and laminated wood of several kinds. The fragrant bark resin is used in perfumery.

15 Ebony *Diospyros celebica*

F. Ébène   G. Ebenholz   I. Ebano   S. Ébano
FAMILY: Ebony (Ebenaceae)   KEY: FI   SOURCE: Celebes Islands, Tropical Asia, Tropical Africa

Fig. 37 Ebony leaves, flower and fruit

Ebony is the symbol of jet-black wood, but most of the hundred-odd ebonies that grow in the tropical forests have heartwood that is not wholly black. Instead, they show bold irregular stripes of bright brown, grey or greenish black on a deep black background. This gives striking decorative effects, and leads to their widespread use as surface veneers. Our specimen is of Macassar ebony, native to the Celebes Islands of Indonesia, which is exported on a considerable scale to America and Japan for slicing into veneers. In solid form, ebony is used for small sculptures by craftsmen, and particularly for the carved heads of dancing-girls which are carved on the island of Bali. Wood of an even colour is naturally chosen for this work.

Ebony has long featured in trade from the East Indies to the Western world, and was used even in Roman times for artistic woodware such as bowls and cabinets. The tendency today is to use the darker, wholly black, material as a contrast to other brightly coloured woods. Large pieces are scarce since it never makes a stout tree, and the pale yellowish-white or pinkish sapwood has to be discarded. Heartwood is strong, very heavy and extremely hard; it holds a natural gum which gives its dark surface a subdued lustre.

Specialized uses include drawing instruments, xylophones, chessmen, castanets, violin mutes and pegs, handles for cutlery

117

or brushes, brush-backs, clothes-buttons, beads, bowls and caskets. Ebony is widely used for making musical instruments, including bagpipes. For all these purposes a hard, strong, attractive material is required, and cost is secondary.

Inevitably, ebony is often imitated by other woods, such as hornbeam, which are stained to look black.

Ebony is a relatively small tree growing in the lower storeys of tropical rain forests. It seldom exceeds 100 feet tall or 6 feet round; the bark is rugged, with squarish plates. Its leaves are simple, oval and pointed, and have a leathery texture, with distinct veins. Small white flowers are borne singly in leaf-axils. Each has a bulbous green calyx and five white petals, which are united below but make a five-pointed star above. The fruit is a small round or egg-shaped structure.

Ebony trees of various species are found in tropical Africa, India, Ceylon, Malaysia and Indonesia.

16 Elm *Ulmus procera*

F. Orme   G. Ulme, Rüster   I. Ulmo   S. Olmo
FAMILY: Elm (Ulmaceae)   KEY: E4
SOURCE: England, Europe, U.S.A., Eastern Canada, Temperate Asia

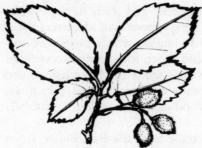

Fig. 38   Leaves and winged seeds
of American elm

Elm is a tall, fast-growing tree of the rich valleys and bottomlands, which demands good soil, with ample moving moisture, and prospers in full sunlight. In the spring the elm cambium forms a ring of large pores, to carry the first rush of sap, but this thirsty tree's demands for water continue during the summer and are met by adding many smaller pores scattered through the wood. These are not set in straight lines, but are ranked in little sideways steps. This produces a very distinctive figure that is evident on every exposed surface of elm wood. It is a pattern of large and small vessels that resembles the feathering of a bird. It has been aptly called 'partridge-breast' figure.

The general run of the wood fibres follows this stepped pattern, and so there are no clear long rays to provide planes of division running out from the centre of the tree. This makes elm, in practice, impossible to cleave or split. It can only be worked with saws or sharp-edged chisels that cut right across its interlocked grain. There are many traditional uses for which 'unsplittability' is a real practical need. One is the seat of a chair, because if legs are driven into this firmly they tend to split it apart. Only elm, of the common hardwoods, stays firm, so it was always chosen for Windsor chair-seats. The hub or nave of a wooden wheel was always made of elm, because the spokes – which were of oak – had to be driven hard into it. The heads of heavy wooden mauls, sledge-hammers and post-drivers are made of elm because it does not shatter on impact. So are boxes and cases to carry tin-plate and heavy metal goods.

Elm trees grow to large sizes and yield broad planks. This makes it handy for the weatherboarding of farm sheds, and for rustic furniture, such as garden seats and tables. But it has little natural durability at ground-level, so will not serve for fencing. It is probably its availability in large sizes, linked to the fact that it lasts indefinitely when placed well below ground-level, that has made it the traditional wood for coffins.

We see elm most frequently as a furniture timber. Sometimes it is cut into intricately patterned slash-grain veneers and sometimes it is sliced to give a bright radial pattern. But it is cheap enough to be used in the solid and this gives a marvellous variety of surface figure. It is strong, firm and stable, promising a century or more of service. Elm is a favourite wood for carvers and turners. Its remarkable grain can be exposed artistically on wood sculpture, and it makes strong, richly grained bowls and platters in sizes not easily achieved with other timbers.

An interesting historic use for elm was as water-pipes and water-pumps, before the days of metal. Great trunks were hollowed out by hand with boring augers, and underground pipe systems were constructed, in London and other large cities, by driving such trunks together. Joints were made by the simple method of pointing the 'male' end of each 'pipe' and forcing it into the larger 'female' end at the butt of the next hollowed-out tree-trunk. Shut away from air, these pipes last indefinitely, and some are still dug up today, after 250 years below ground. The barrels and working buckets and valves of pumps once used in country villages to raise water from wells were also shaped from elm. The same designs were later copied in iron.

Elm sapwood – seldom seen on manufactured timber, is white in colour and forms only a thin outer zone. Heartwood has a warm brown colour. Freshly worked elm has a distinctive earthy smell, like that of rich garden mould.

Four kinds of elm, and several local varieties, are native to the eastern and mid-western States. Three of these, the American elm, *Ulmus americana*, the rock elm, *U. thomasii*, and the slippery elm, *U. rubra*, thrive as far north as Ontario. Winged elm, *U. alata*, grows only in the south-east. Europe likewise has many species, the most distinctive being the wych or mountain elm, also called the Scots elm, *U. glabra*. This is a northerly or mountainside tree, which yields a tough timber but rarely forms a large undivided trunk; it is apt to branch too low to suit the timber merchant. In many European and Asiatic countries elm leaves are valued as fodder for sheep, goats, horses and cattle, and trees are regularly lopped to secure this.

In England few natural elmwoods remain, for the good land they occupied was cleared for ploughing centuries ago. But for the past 300 years landowners have planted and maintained elms along hedgerows, where they thrive in the good soil and full sunlight. There are several regional kinds, which renew themselves, if felled, by vigorous sucker-shoots that spring up from their roots – though the wych elm will not do this. The finest is the English field elm, *U. procera*, a magnificent tree that grows rapidly to a height exceeding 140 feet, with a trunk up to 31 feet round. Its foliage forms huge billowing masses and its unique contribution to the lowland scene has been well expressed by painters such as Constable. The finest elms grow in Suffolk and neighbouring East Midland counties.

During this present century elms in Holland, neighbouring European countries and parts of eastern North America have suffered very severely from an elm disease that was first noticed in Holland about 1925. This is caused by a fungus but is carried from tree to tree by small bark-boring beetles. Luckily some species and strains of elm are immune, and new planting is now done with these.

Elm leaves are simple in shape and oval in outline, with a pointed tip and a toothed edge. A remarkable feature makes them easy to name, in that the base of every leaf is always uneven or oblique; the two sides never match. Elm flowers open late in the winter, about February, well ahead of the leaves. They form purplish-red, catkin-like clusters on bare twigs. Each flower develops a single small seed, set in the centre of a large, yellowish-green wing. This seed ripens rapidly, and is dispersed by the wind soon after the leaves open in spring. Much seed is infertile, but sound seeds quickly sprout after reaching moist soil.

17 Eucalyptus *Eucalyptus regnans*

F. Eucalyptus   G. Fieberbaum   I. Eucalitto   S. Eucalipto
FAMILY: Myrtle (Myrtaceae)   KEY: E5
SOURCE: South-eastern Australia, California (introduced), Sub-tropics generally (introduced)

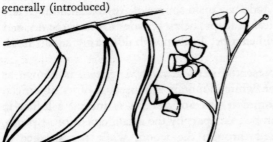

Fig. 39 Eucalyptus: leaves and seed-pods

Eucalyptus trees, sometimes called 'eucalypts', are native only to Australia and Tasmania, though several kinds are now widely grown for timber and ornament in many other countries with a similar climate, including California, Spain, South America, India and South Africa. Most of them have similar foliage, flowers and seeds, but wood and bark can vary a great deal, and so can their general vigour. A typical eucalypt bears two sorts of leaf: young plants and young stems have rounded 'juvenile' leaves in opposite pairs, clasping the stem; older plants and twigs bear solitary 'adult' leaves on distinct stalks, and the shape depends on the species of tree. Each adult leaf adjusts itself on the stalk so that it presents its edge, and not its face, to the sun. This allows the sun's rays to strike through the foliage, and there is little shade below a eucalyptus tree. It is a device to lessen water loss during the hot Australian summer. Eucalypts grow under a 'Mediterranean' climate with warm wet winters and hot dry summers; they are evergreen, in order to make best use of this difficult sun and water régime.

Another device to lessen water loss is the waxy bloom on the surface of the foliage, which gives it a bluish sheen. Associated with this is the strongly smelling eucalyptus oil, which forms in glands throughout the leaves. This is obtained commercially by distilling large quantities of fresh eucalyptus leaves with steam. It is widely used as a cure for colds, and in perfumery. It is both volatile and highly inflammable, and its presence causes eucalyptus trees to blaze with exceptional ferocity in Australian bush fires.

Eucalyptus bark is steely blue in colour, but peels away continually in irregular strips, exposing yellowish bark below. Hence the tree-trunks always present a dappled appearance.

The flowers are yellowish white and appear in clusters. They have many stamens, which give them the general appearance of little feathery tufts or powder-puffs. Being pollinated by bees, they bear ample nectar and are a main source of Australian honey. The seed-pods are conical in shape, being broader towards the tip, and bluish in colour when ripe. Each pod holds many small, hard, black seeds.

The timber represented here, *Eucalyptus regnans*, is known as 'Tasmanian oak' or 'white mountain ash' because of its superficial resemblance to common oak and ash. It is indeed ash-like in appearance but can be told apart by the arrangement of its pores, which are scattered through the wood, while the absence of distinct rays marks it out from oak. It is hard, heavy and strong, and can well take the place of oak or ash in constructional work or as solid furniture. It has long been so used in Australia, and is also exported, both in solid form and as decorative veneer.

The tree that yields it is the tallest in Australia, and may scale 300 feet tall by 30 feet round. Its bark is rather stringy in character, which aids identification when the foliage is high out of reach. Its adult leaves are remarkably slender, and are sickle-shaped. They are a favourite food of the koala, or tree-bear, a quaint beast that resembles a cuddly teddy-bear and carries its young in a pouch, like a kangaroo.

Eucalyptus regnans forms large pure stands in the uplands of New South Wales, Victoria and Tasmania. Much of the finest timber was felled by settlers to clear land for agriculture.

Other species of eucalypts are planted in sub-tropical lands as sources of fuelwood and pulpwood for paper-making. When cut back, they sprout again, and so yield perpetual crops. Unfortunately, all the eucalypts – even those that grow on snowy Australian mountains – are frost-tender, and none succeeds in northern North America, nor in Europe north of the Alps.

18 Iroko *Chlorophora excelsa*

FAMILY: Mulberry (Moraceae) KEY: E6
SOURCE: West Africa

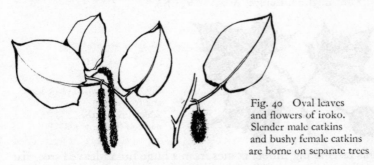

Fig. 40 Oval leaves and flowers of iroko. Slender male catkins and bushy female catkins are borne on separate trees

Iroko is a strong, mid-brown, durable timber that has a close resemblance to teak, indeed it is often marketed as African or Nigerian teak. It can readily be distinguished by its scattered pores, contrasting with the ring-porous structure of true teak. Yellow bands of soft tissue form a lively zigzag pattern on all surfaces, and this, too, contrasts with the uniform colour of teak. Iroko is fairly soft and easy to work, and the Africans use it for carving wooden platters and bowls, spoons and similar household utensils, and for sculpturing wooden figures and statues. Canoes are hollowed out of large trunks, and because of its natural durability it is used for fence posts and building construction.

Supplies are plentiful, and much iroko is exported as large logs or baulks. In America and Europe it is employed for much the same jobs as teak, including benches for workshops and laboratories, counters for shops, sturdy joinery, outdoor seats and tables, and in boat building.

Iroko grows wild as a tall savanna tree right across Africa from Sierra Leone through Ghana and Nigeria to Tanzania. Its bark is greyish white, wrinkled and scaly, and holds a sticky latex which oozes out from any wound and seals it off. The young twigs, purplish in colour, bear simple oval leaves that end in blunt points. The flowers are catkin-like, and droop downwards. Native names include 'odoum', 'mvule' and 'kampala'.

19 Lacewood Plane *Platanus acerifolia*

F. Platane, Bois puant   G. Platane   I. Platano   S. Plátano
FAMILY: Plane (Platanaceae)   KEY: B3
SOURCE: England, Europe, Asia Minor, USA

Fig. 41 Lacewood plane leaf
· and fruit-catkin,
which gives it the
name of 'buttonwood'

This fascinating timber comes from a huge broad-leaved tree that bears many names. It is called 'American sycamore' because of the resemblance of its leaf to that of the English sycamore maple (see p. 151), and 'buttonwood' or 'button-ball' tree' because its peculiar fruit-mass looks like an old-fashioned leather button. In England it is called 'plane', while other European names – *Platane* in Germany and France, *platano* in Italy and Spain – are based on the old Latin *platanus* and Greek *platanos*, all meaning 'broad-leaved'. Why lacewood? As usually cut as ornamental veneer, plane-tree wood shows its broad rays surrounded by curving bands of other tissues and the whole effect is that of a lace fabric. This cutting is done radially, along the rays which run out from the heart of the tree.

The American plane tree or sycamore, *Platanus occidentalis*, flourishes in all the eastern and mid-western States, and grows larger than any other broad-leaved tree. The size record is held by a veteran in Indiana, 168 feet tall by 33 feet round. American sycamore was introduced to England about 1600, but made little headway as the climate proved too cold. About the same time another plane tree, *P. orientalis*, was brought in from Turkey, and this, too, grew indifferently in the cool moist climate of the British Isles. The two were planted side by side in the University Botanic Gardens at Oxford, and in about 1670 a lucky international marriage took place which produced the so-called London plane, *P. acerifolia*. This wonderful tree shows great hybrid vigour and tolerance of the adverse climate that discouraged its parents from settling in Europe. It is readily increased by cuttings and has become the leading shade tree in London, Paris, Brussels and most other large cities of Europe. One of its great advantages is that it takes kindly to the repeated prunings required to restrain its rapid growth.

Plane trees are easily known by quite unusual features. Their bark is smooth and greyish brown, but continually flakes off in patches which expose pale yellow younger bark below. This gives the trunk a dappled pattern, as though the sunshine were always rippling over it. It also helps the plane to thrive in smoky cities, for as soon as old bark gets its breathing pores clogged with soot, it is shed; a tree must breathe oxygen through its bark, and also its roots, as well as through its leaves.

The leaves of the plane are broad and five-lobed, like those of the maples. But two points distinguish them clearly – they are always placed *alternately* on the twigs (never opposite) and the base of each leaf-stalk is hollowed out into a deep cone – like a dunce's cap. The winter buds, which are of course alternate also, are neat little reddish-brown cones, with only one outer scale in view. The scar of the previous year's leaf-stalk completely encircles this bud – a unique feature.

The 'button-ball' flowers and fruits are very odd. Each tree bears circular flower-masses hanging down in twos or threes at intervals on long stalks. All look alike, but on the same tree some will consist entirely of male flowers, others entirely of female ones. The individual flowers of each sex are very small and greenish in colour, without petals, scent or nectar, and when they open in June the pollen is carried by the wind. Male flowers then fall, while the female ones ripen to brown seed-masses that hang on the leafless trees as conspicuous 'bobbles' right through the winter. In spring they disintegrate into scores of tiny seeds, each carrying a tuft of yellow hairs, which help to carry the seeds on the winds. Some people find these dry, dusty, hairy seeds highly irritating. Much seed is infertile, but good grains sprout by opening two curious sickle-shaped seed-leaves, followed by simple, unlobed juvenile leaves before the broadly lobed adult foliage appears. Later growth is rapid, and the American plane tree holds its own well in fertile bottomland forests, despite competition from other trees.

The shade trees in the squares and boulevards of European cities are cherished, and it is rare for timber to reach the market from such sources. Most is felled in America, south-east Europe or Asia Minor, where plane trees form natural forests. The timber is hard, strong, moderately heavy and stable, and proves satisfactory for furniture, joinery, and a wide range of wooden objects like brush-backs and small tool-handles. Its pale brown surface usually shows a yellowish-pink tinge, and there is no obvious distinction between sapwood and heartwood. Out of doors it has no natural durability, but it meets a wide range of indoor uses and is also a good firewood. Large logs are sliced radially to yield the characteristic lacewood veneer which is widely employed in high-class cabinet-work.

20 Lime *Tilia europaea*

F. Tilleul   G. Linde   I. Tiglio   S. Tilo
FAMILY: Lime (Tiliaceae)   KEY: B4
SOURCE: Europe, USA, Eastern Canada

Fig. 42   Leaves and fruit stalk of American lime, with typical papery bract

When the lime tree is grown in gardens it is often called the 'linden tree', from its Old German name of *Linde* which has survived from Anglo-Saxon days. In forests and timberyards, however, it is usually called 'basswood', a name derived from its remarkably thick and very tough inner bark or bast.

The leading American species, which provides most of the timber in commerce, is American basswood, *Tilia americana*, found in all the northern States from New England west to Dakota and south to Oklahoma and Arkansas; it also grows in Quebec and Ontario. The white basswood or beetree linden, *T. heterophylla*, has a more southerly range from the Appalachians down to Florida. American limes are also cultivated in gardens and parks and as a street shade tree, but the commonest cultivated lime tree is European lime, *T. europaea*. This probably arose through the chance interbreeding of two wild species, the broad-leaved lime, *T. platyphyllos*, and the small-leaved lime, *T. cordata*. Both grow wild in the lowland forests of western and Central Europe, including England and Wales. But they are rather rare, because their foliage proves highly palatable to livestock.

The lime-tree leaf has the shape of a conventional heart, sometimes perfectly symmetrical and sometimes rather uneven at the base. Winter buds, set alternately along slender twigs, are reddish brown and always show two outer scales, like the finger and thumb of a hand. Lime flowers can be known at once by a remarkable feature – an oblong papery bract fixed to the main flower-stalk. This stalk bears four or five smaller ones, each carrying a single yellowish-white flower. Every flower has five sepals, five petals, many stamens, a single ovary and nectaries that attract honey-bees. They open in June, and ripen by October into hard, round one-seeded pods. When these fall, the papery bract acts as a wing and helps their spread on the wind. Lime seedlings, which spring up next spring, have odd-looking seed-leaves, each being split into several 'fingers' like those of a human hand.

Beekeepers find that lime blossom is a rich source of nectar, yielding honey that holds the pleasing fragrance of the flowers. But it is sometimes tainted by honeydew, a sweet sticky substance secreted by aphids that feed on lime leaves. Honeydew turns black after falling, owing to the action of a fungal mould. In some summers this black sticky substance coats pavements, car roofs and windscreens beneath lime trees, making lime a less-than-perfect choice for street planting. Otherwise lime is a lovely tree that looks remarkably attractive spaced along avenues.

Its bark was long used as a cheap textile fibre. The Russians made it into ropes and mats, and gardeners still use it to tie up bundles of plants. Externally it is smooth and grey. Fibre is secured by debarking logs and beating sheets of bark with mallets to remove the outer surface from the tough interior layers.

Lime timber is an even pale yellow in colour, without distinction of heartwood and sapwood. It shows scarcely any figure. Rings and rays are obscure, pores diffuse and very fine. It is soft, light in weight, and has a pleasant smell when fresh-cut. Lime has a pleasing natural lustre and is soft enough to be easily carved, yet firm enough to hold a precisely cut surface well. Once seasoned it is very stable, free from both shrinkage and warping. These properties make it the favourite wood for sculptors, carvers and pattern-makers. Grinling Gibbons and other masters of wood-carving found it ideal for delicate sculpture. Makers of hat-blocks, shoe-trees and engineering patterns prefer it because it retains shapes and sizes well. It is used for piano-keys and other precisely made parts of musical instruments for the same reasons.

Lime takes stains well and has good bending properties. It is not naturally durable, and is seldom used out of doors.

21 Mahogany, African *Khaya ivorensis*

F. Acajou d'Afrique   G. Khaya-Mahagoni   I. Mogano Africano
S. Caoba Africano
FAMILY: Neem tree (Meliaceae)   KEY: D5   SOURCE: West Africa

Fig. 43   Flower and compound leaf of African mahogany

The value, and at times the scarcity, of Honduras mahogany (see No. 22) have led timber-importers to seek similar woods in other lands. Africa has proved the most productive source, and the botanical genus *Khaya*, which is closely related to the true mahogany genus *Swietenia*, yields acceptable substitutes as large-sized logs and in considerable quantities. Several species grow in the rain forest belt along the West African coast, from Sierra Leone east to the Congo, and all alike are marketed as African mahogany.

The species described here, *Khaya ivorensis*, is one of the largest jungle trees, often 150 feet tall with a cylindrical bole that may become clear of branches for 90 feet up. The trunk may be 18 feet round, measured above the great basal buttresses that can run 15 feet up from the ground. The evergreen, leathery leaves are compound, with about six pairs of oval leaflets. The flowers, borne in clusters, are yellow. The fruit is a round pod, about three inches across, which holds about fifty hard winged seeds. The outer bark is thick and rough and is shed in circular patches; there is a bright red inner bark.

In appearance and working properties African mahogany is a close match for the nearly related Honduras kind. It is generally held to be a little less stable, and is rarely so beautifully figured, but it has been applied with success to the same range of uses. The two timbers are far from easy to distinguish, but a useful pointer is that the pores or vessels of African mahogany are usually grouped in clusters, rather than being evenly spread. Sapwood, when present, is pale yellow.

African mahogany is also known as 'Benin', 'Grand Bassam' or 'Lagos mahogany' from these ports of shipment. In America it is often marketed under its botanical name of 'Khaya'.

22 Mahogany, Honduras *Swietenia macrophylla*

F. Acajou Amérique   G. Mahagoni, Echtes Mahagoni
I. Mogano Honduras   S. Caoba
FAMILY: Neem tree (Meliaceae)   KEY: D6   SOURCE: Central America

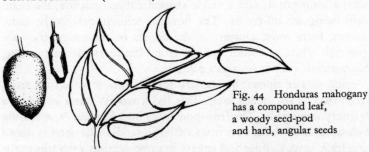

Fig. 44 Honduras mahogany has a compound leaf, a woody seed-pod and hard, angular seeds

As soon as the first Spanish explorers reached the West Indies they began to examine the timber resources of their New World. They needed wood to repair their ships, battered by the long Atlantic crossing, and they also sought something valuable to load into their holds as a return cargo. They soon discovered the wonderful red-brown timber that they called 'caoba', which we know as 'mahogany'. As early as 1514, it was being used in the island of Santo Domingo. Cortes, the conqueror of Mexico, used it for shipbuilding. By 1584 Philip II of Spain was employing it to decorate the Escorial Palace in Madrid. In England, the first recorded use was at Nottingham Castle in 1680. From 1715 onwards regular shipments took place and it was widely adopted for high-class tables, chairs, chests, bedsteads and furniture generally.

At that time it was easy to get the large planks the cabinet-makers preferred from Cuba, Jamaica and other West Indian islands. As large trees were cut out, the loggers turned to Honduras, on the Central American mainland, which now gives its name to all mahogany cut in this region. During the present century shortages in Central America have made exporters look to South America for supplies, and much mahogany, of this or closely allied species, is now felled in Colombia, Venezuela and along the upper reaches of the River Amazon in Peru, Bolivia and Brazil. Plantations have been established in Trinidad, India and British Honduras.

Mahogany is one of the largest trees of the tropical American rain forest. It frequently exceeds 100 feet in height and its stout bole may measure 40 feet round, above the large basal buttresses. It is usually felled from platforms built 6 feet above the ground, so that the timbermen can clear these swellings. It is an evergreen, with curious compound leaves, about nine inches long, composed of three to four pairs of leaflets. Each leaflet is an extended oval, with a long point, and a sickle-shaped curved outline, the main vein being set off-centre. The flowers, which open in the rainy season, form loose clusters in the angles of the twigs; they are greenish white in colour, on the true Honduras mahogany tree, but reddish yellow on related kinds.

Each flower ripens, by the dry season, into a large oval fruit, from two to six inches long. This has a woody outer coat and a leathery inner pulp. The fruit-pod splits into five leaves, revealing below each of them two rows of large seeds. The seed is about one inch long, slender and square in cross-section with ribs at the angles, and bears an oblong wing.

When freshly felled the heartwood of mahogany is bright pink, with a thin zone of colourless sapwood. On exposure to the sunlight, the heartwood rapidly darkens to the rich coppery-red shade that we know so well. Mahogany is rather a light timber and also quite soft. Its great virtues are ease of working and stability; once shaped it does not shrink or warp. It is naturally durable, and has long been favoured by shipbuilders, from the Amazonian Indians, who carved it into canoes, to the modern designers of luxury yachts.

Mahogany finds its main employment in the furniture trade, where it is valued for colour, workability, general stability, and the fact that it is available in large sizes. It is also used for high-grade joinery, shop-fittings, showcases, counters and interior decoration. The most attractively figured logs are sliced into veneers, showing wavy or curly grain. Mahogany is readily polished to highly lustrous surfaces. Though fashions change, it has held its place as the leading furniture wood for four centuries.

Honduras mahogany shows no marked structural features. It is an even deep coppery-brown colour, with obscure rings and only slight grain, but its rays can be seen clearly on radial-cut surfaces as smooth plates which reflect the light. The pores, usually stained or discoloured darker, are large and evenly spread through the wood.

In Mexico mahogany is commonly called 'zopilote' or 'chiculte'; only rarely is it given its full name of 'zopilozontecomacuahitl'.

23 Maple *Acer saccharum*

F. Erable à sucre   G. Zuckerahorn   I. Accro da zucchero
S. Arce de azúcar
FAMILY: Maple (Accraceae)   KEY: A5   SOURCE: New England, Eastern Canada

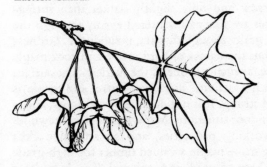

Fig. 45 Maple leaf
and winged fruits

Maples of many kinds grow in temperate forests right across North America, Europe and Asia. Best known is the sugar maple, also called 'white maple', 'hard maple' or 'rock maple', which is native to the north-eastern United States and south-east Canada. All the maples bear their leaves in opposite pairs, which means that their winter buds are also paired and opposite, except for the terminal bud that prolongs the growth of each shoot. In most sorts of maple the leaves are lobed, and each lobe ends in a distinct point. This gives the outline that is so familiar on the maple-leaf flag of Canada.

Maples are renowned for the brilliant tints of orange, scarlet and gold that their leaves assume in the autumn of the year – one of the most magnificent sights of the North American forests.

Maple flowers are borne in early spring, in clusters, and though flowers of both sexes are found in each bunch, individual blossoms are often wholly male or wholly female. A perfect flower holds five sepals, five petals, five stamens, an ovary and nectaries to reward honey-bees. The fruit is most distinctive, for each

131

fertile female flower produces two hard round seeds, facing one another. Each seed has an oval wing, and when they fall, either singly or united, they spin round in the air like the blades of a helicopter. Maples seed abundantly every fall, and numerous seedlings sprout on the forest floor next spring. Each seedling has two oblong seed-leaves, which are followed by oval, singly-pointed juvenile leaves before the adult lobed leaves appear.

Bark varies with species; in the sugar maple it is thin and broken into small squares by shallow ribs. The largest recorded sugar maple, growing near Bethany, West Virginia, is 110 feet tall and 20 feet round.

Maple timber is pale cream or biscuit-brown to yellowish-white in colour, without distinction of heartwood and sapwood. Summerwood is narrow and only slightly darker than spring-wood, while the pores are fine and scattered evenly through the timber; therefore the grain, though distinct, is subdued. Hardness varies with species, but the names of hard maple and rock maple remind us that this sugar maple is one of the hardest. The surface is always smooth with a definite, attractive lustre; and maple is firm, fairly heavy and strong. It is not durable out of doors, and is usually employed in indoor situations. Maple is first-class firewood.

Strength, good working properties, and an attractive clear pale-coloured surface make maple a valued timber for high-grade furniture, and it has been so used since colonial days. It is also a very good joinery timber, for all kinds of shelves, fittings, benches and counters, and much is employed in boatbuilding and ships' equipment. It is the best of all woods for flooring – either as narrow boards or parquet blocks. Dance-halls are always floored with maple because it wears slowly, smoothly and evenly, without splintering. Wood-sculptors and turners like it because, though hard, it works to a smooth finish and remains very stable.

Rather surprisingly, a good deal is used for paper-pulp. When it is added to a pulp made from softwood fibres, maple adds strength and solidity and results in an excellent writing- or printing-paper.

Sugar maple is the source of maple syrup and maple sugar, both obtained by concentrating its sweet sap. This can be secured only in the spring, during a spell of about three weeks, just as the snow is thawing. In the winter every broad-leaved tree that loses its leaves is obliged to store its food reserves in its trunk, branches and root-wood, in the solid, insoluble form of starches. Much is stored in the rays. When food material is needed for the fresh growth of leaves and shoots, it is converted into soluble sugar for ease of transport upwards to the twigs and buds. Although many people think that sap 'goes down' in the winter, it is actually present in the tree-trunk and stems all round the year, but it is static and will not flow out. Once the earth warms up, the tree's roots become very active and send a steady stream of sap coursing upwards towards the branches. If a maple tree is wounded at this time of year, it will 'bleed' for two or three weeks. All broad-leaved trees mobilize sugar sap in spring, but only a few will bleed.

The Indians of North America discovered long ago that the sugar-rich sap of the maple tree was exceptionally sweet and pleasant to drink. But they could not make syrup or candy because they had no iron pans. The sweet juice must be concentrated by boiling within a few hours of harvesting. Otherwise it is quickly attacked by bacteria and turns sour. The colonists quickly discovered this, and for several generations they used the maple tree as their main source of everyday sweetenings. Eventually it was displaced by cheaper imported cane sugar, but it remains popular as a luxury sweetmeat because of its delicious nutty flavour.

Most 'sugar bushes', as they are called, are woodlots owned by small farmers in New England and south-eastern Canada, who can find the labour, usually from their own families, to gather in the harvest during the brief season. In the traditional method, a small metal spout is hammered into each tree, and a cup is hung beneath it. Every day each tree is visited, and a cupful of sap is poured into a bucket, producing at length a heavy load that is carried on a sledge over soft thawing soil to the boiling-shed. There the sap is concentrated in a large iron pan over an open fire fed with waste maple branchwood. It may be marketed as a thick, rich, brown syrup, or made to crystallize out as soft candy from a stronger liquid obtained by longer boiling.

Technologists have been busy here as elsewhere, and today much sap is collected through plastic tubes running from several trees to a plastic container. It is later concentrated in a vacuum pan, the whole system being planned to give consistent products with less labour.

Sugar bushes stand up to repeated tapping year after year, provided they are properly maintained by thinning out unwanted trees and encouraging the growth of saplings as the older trees lose vigour. Man is gathering in only a fraction of the food material that the maple's leaves win from the air around them, every summer, through the magical chemistry that uses the sun's energy as its mainspring.

24 Oak *Quercus robur*

F. Chêne   G. Eiche   I. Rovere   S. Roble
FAMILY: Beech (Fagaceae)   KEY: B5
SOURCE: England, Europe, USA, Eastern Canada, Japan

Fig. 46   Leaves and acorn of English oak, *Quercus robur*

Oaks of many kinds flourish in the woods of North America, Europe and northern Asia. Features common to them all include the characteristic acorn, a hard round seed carried on a round cup, and the grouping of numerous winter buds close to the tip of each twig. This is a useful point, since it aids identification in winter. Leaf shape varies a great deal, but the leaves are always placed separately on the twigs, never in opposite pairs. In most oaks the outline is irregular, with a series of lobes and bays breaking the oblong shape.

Oak flowers rarely attract notice, since they are greenish yellow and appear after the leaves. Male catkins are bunches of long stalks that bear separate small flowers at intervals; the anthers scatter pollen, and then the whole catkin falls. Female catkins are even smaller, mere groups of tiny flask-shaped green flowers. After pollination they ripen rapidly, usually producing the acorn in four months, though in a few species it needs sixteen months to ripen. The woody cup is formed from numerous little leaves, or bracts, that become fused together and hard. Oak bark, though smooth on young stems, soon becomes thick and rugged; it is mid-grey in colour.

The acorn crop – like the apple harvest – varies from one year to another. In a good year, or 'mast' year, it is so abundant that many acorns escape being eaten and survive through the winter to sprout next spring as seedlings. Acorns have many enemies, including squirrels, wood mice, game birds, pigeons and crows. Though humans cannot digest them, they make excellent food for pigs. When William the Conqueror made the 'Domesday Book', a taxation roll for English villages, in 1086, he valued the woodlands according to the number of swine they could support.

Most parts of an oak tree are full of tannins or tannic acids – powerful chemicals that have the remarkable property of making skins and hides resistant to decay. Before synthetic chemicals or alternative plant extracts had been discovered, oak bark was the best source of tannin for making leather. Whenever oaks were felled, bark was carefully stripped and stacked, with its outer surface upwards, to dry. Later, at the tannery, it was steeped in water to release the tannin. Oak bark fetched high prices and was often worth more than the timber! For centuries it was regularly harvested in Europe and, later, in North America.

Tannins in the wood make the heartwood exceptionally durable out of doors, even without treatment, though the sapwood proves perishable. But they react with iron to cause inkstains, which have in fact the same chemical composition as writing-ink. These stains are ignored in rough outdoor work; for ornamental construction indoors, however, oak must always be secured with wooden pegs, brass screws or other metals holding no iron. It is, in any event, so hard that ordinary nails cannot be driven into it. A strong odour of tannic acid arises from every newly worked surface of fresh oak, making it unmistakable.

Oak timber has strongly marked features that make it simple to identify. It is ring-porous, and the circles of large pores make the softer, less dense springwood of each annual ring stand out clearly from the harder and denser summerwood. The pores show as deep striations or vessel lines on longitudinal surfaces.

The rays are well developed. On the end-grain they can be seen as clear straight lines, and they are also visible on a slash-sawn or tangential surface. Sawing or slicing an oak log radially, from the centre to the circumference, reveals the rays as broad plates, up to one or two inches deep and almost equally wide. They appear definitely harder and smoother than the rest of the wood and reflect the light so strongly that radial-cut oak can fairly be regarded as 'two-coloured'. These rays make the famous 'silver grain' of oak, rightly valued as an ornamental feature. Much oak is quarter-sawn to display it.

Except for this ray figure, most oak is yellowish-brown when freshly cut. It is, however, readily stained and may be found in use in a variety of shades and finishes, such as the artificially whitened limed oak or bleached oak. Brown oak, discussed later, results from natural staining by a fungus. Oak sapwood is always pale, from yellowish-brown to whitish-yellow in colour.

White oak, *Quercus alba*, is the leading North American species. Europe has two good timber oaks, distinct in the textbooks but not so in the forests, since they interbreed. Pedunculate oak, *Q. robur*, has stalked acorns, while sessile oak, *Q. petraea*, has stalkless acorns carried directly on the twigs. Japanese oak, occasionally exported to Europe and America, is cut from *Q. mongolica* variety *grosseserrata*. In appearance and working properties all these four kinds are much alike, and are used for similar purposes. But the North American red-leaved oaks, such as *Q. borealis*, are less strong and must be limited to less exacting jobs.

Oak grows slowly, but it is a remarkably adaptable tree, thriving on a wide range of soils under varied climates. It springs up readily from self-sown seeds, and is also easy to propagate in nurseries for forest planting. Much good oak-forest land has been cleared for farming, both in Europe and in America. Yet despite centuries of harvesting, oak remains a major timber, with good supplies in prospect. European oaks may reach great girths, up to 44 feet round, and attain ages exceeding 1,000 years; but they are not tall trees, and the greatest recorded height is 128 feet. American white oak, which has a natural range over all the States east of Kansas and just reaches Ontario and Quebec, makes a taller tree, up to 150 feet tall; its greatest recorded girth is only 30 feet, however, and its greatest age around 600 years.

Oak was for long the standard, in fact almost the only, building timber in western Europe and the eastern United States. It can readily be cleft or hewn, by methods described on page 9. It is naturally durable and remarkably strong, and serves equally well for posts or beams, clapboard or roofing shingles, or even the rafters of a vast cathedral. Kept reasonably dry, it can and does endure for centuries. It was also the main shipbuilding timber, providing naturally curved, immensely strong 'crooks' and 'knees' for framing ships' hulls and supporting decks. The Anglo-Saxons crossed from the Rhine to England in open boats built with cleft-oak planking, and the Pilgrim Fathers, a thousand years later, sailed over the Atlantic in the oak-framed *Mayflower*.

Around the homestead, cleft and hewn oak were used for farm fencing, wheel spokes for horse-drawn carts, and ladder-rungs, because they were strong enough to take all likely strains. Oaken chests, often constructed of hand-adzed boards, were used to hold valuables, and men sat on oak benches to eat off oak tables. Hand-made chairs and dressers surviving from those days are highly valued for their simple integrity of design and workmanship.

Modern technology has ousted oak from many traditional uses, replacing it with metals, concrete, softwoods or even plastics. But none of these materials share its beauty of grain, and it is still employed where appearance counts. Much is used in high-class joinery and fittings, but the biggest demand is for sound, substantial furniture. Fashions vary, but oak always returns to popularity on its proven worth. Every surface is different, and full of character, and this rewarding timber is cheap enough to use in the solid form.

It will also, if required, give attractive and serviceable veneers. Some of these make effective use of the silver grain, others are skilfully sliced to reveal burr or crown figure.

Coopers use oak for making wooden barrels, and though these are now seldom used for beer they are essential for the proper maturing of choice wines, sherry, port, brandy and Scotch whisky. Barrel-staves for spirits must always be radially sawn or hand-cleft, since it is only the ray tissues that stop the alcohol from seeping through the wood.

Other uses of this sturdy, versatile timber are in mining, bridge-building, wagon construction, dock and harbour work, and every engineering use that demands exceptionally strong timber. Door-steps, window-sills, and other elements of buildings that are exposed to much weather and wear are still made from oak, and so are greenhouses and garden-sheds.

Minor uses include paper-pulp, hardboard and other forms of man-made board. Branchwood is used on a large scale for fire-wood and also, even today, to make charcoal. Oak charcoal was once the main fuel for iron-smelting, before coke came into use, and is nowadays applied instead to a variety of chemical reactions in industry. It holds the structure of the wood so well that charcoal recovered from Roman sites in Kent has been conclusively identified as originating in a primeval English oak forest, 1,600 years ago.

25 Oak, Brown *Quercus robur*

F. Chêne pollard   G. Traubeneiche   I. Rovero pollardo
S. Roble pollardo
FAMILY: Beech (Fagaceae)   KEY: E7   SOURCE: England, Europe

Fig. 47 Sporophores of
the beef-steak fungus,
Fistulina hepatica, which
stains oak brown

Brown oak is also called 'pollard oak', because it is usually cut, as a sliced veneer, from an open-grown tree that has been lopped or pollarded in the past. Pollarding was once a common practice in Europe, where peasant farmers depended on trees growing in pastures for a steady supply of firewood and small poles. If they lopped a tree at 6 feet up from the ground, the sprouts that grew out from the pollarded trunk were safe from attack by browsing cattle, sheep, horses and goats. Therefore they were able to come back about ten years later and harvest another crop of poles. Many pollard oaks are now several centuries old, and since lopping is now rarely done, their branches have grown very stout. The interwoven pattern of grain, from several stems leaving the trunk at one point, can prove very attractive when exposed to view.

But it is the natural staining that gives brown oak its unique appearance, and this is caused by a fungus known as *Fistulina hepatica*. It is also called the 'beef-steak fungus' because its large sporophore – also known as a 'bracket' or 'conk' – is quite good to eat and has a flavour recalling beef. The fungus gains entry to the wood through an open wound that cuts through both bark and sapwood, so exposing its heartwood. Such wounds are of course bound to occur when a stout branch is lopped. It develops unseen by means of thin threads called 'hyphae', visible only under the microscope, which pervade the heartwood, gradually involving the whole trunk. Unlike many fungi that cause serious decay and eventually kill the tree, the beef-steak fungus simply extracts enough nourishment for itself; it also stains the wood owing to chemical changes that occur. After growing for several years it produces its spore-bearing brackets on the side of the trunk.

These are conspicuous structures, often a foot across, and may be exceptionally heavy, weighing as much as 30 lb. They grow outwards rapidly in autumn, but wither away within a month. Their smooth upper surface is liver coloured, purplish-red, blood-red or chocolate. The under surface is made up of countless pale pink pores or tubes which release millions of spores on the air. If the flesh of the bracket be cut across, it is seen to be reddish in colour, while a red juice oozes out. Gourmets gather this odd fungus just as it ripens and becomes tender, losing its earlier sharpish taste; they cook it and eat it like a mushroom.

If, however, a suitable pollard tree bearing beef-steak fungus is spotted by a 'talent scout' from a veneer firm, an offer will be made to its fortunate owner. The wood is very valuable when skilfully converted into decorative veneer, and high prices can reasonably be asked. In all its main characters, brown oak resembles ordinary oak, described above, but the natural staining gives attractively varied colours that cannot be matched by any artificial process.

Brown oak should not be confused with 'bog oak', which is timber that has lain for centuries in a peat-bog, usually in Ireland. The acids of the peat preserve it, but also stain it a uniform dull black. It is interesting as a curiosity, but has little ornamental value.

26 Padouk, Andaman *Pterocarpus dalbergioides*

F. Padouk d'Andaman   G. Andaman Padouk
I. Paduk delle Andamane   S. Paduk de Andaman
FAMILY: Sweet pea (Leguminoseae)   KEY: CI
SOURCE: Andaman Islands

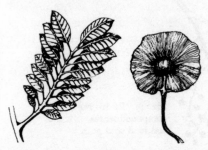

Fig. 48   Andaman padouk bears large compound leaves and winged seed-pods

Andaman padouk is a beautifully coloured wood, showing dark chocolate-brown over a rich purplish or bright crimson background. Its weight, hardness and general feel of great strength and solidity also mark it out. Figured material is widely used as veneer, while solid stock is employed in exacting work where both strength and exotic appearance count. This includes parquet flooring, shop- and bank-counters, railings and banisters, billiard-tables, billiard-cues, panelling, substantial furniture, boatbuilding and ships' fittings. The sapwood, rarely seen on exported logs, is narrow and yellowish-grey.

This remarkable timber grows only in the Andaman Islands, which are situated midway between India and Malaya in the Indian Ocean. There it forms a tall tree, around 100 feet high by 8 feet round, with ascending branches, which thrives along the banks of creeks. During the dry season it stands leafless for two months – a climatic feature recorded by the clear annual rings. About seven leaflets, each about two and a half inches long by one and a half inches wide, make up the large compound leaf, seven inches long overall. The flowers resemble those of a sweet pea, and are followed by winged seed-pods.

At one time this timber was harvested solely by convict labour. The British established penal colonies on the Andamans, and lumbering provided suitable employment for prisoners sent from India and Burma. Padouk is a Burmese word, and similar woods grow in that country. In America, it is sometimes called 'vermilion wood', while the Andamanese name is 'chalanga-da'.

27 Paldao *Dracontomelum dao*

FAMILY: Mango (Anacardiaceae)   KEY: F2
SOURCE: Philippine Islands

Fig. 49 Paldao has large compound leaves and hard seed-pods

Paldao wood shows a wonderfully variegated pattern with grey as its main colour. An amazing irregular mottling of greenish-brown, dark brown, dark grey or black streaks runs over a yellowish-grey or pinkish-grey ground. No two surfaces are alike, and this lively figure makes paldao a favourite timber for the veneered finish of cabinets, radios and occasional tables. There is a remote resemblance to European walnut, and this, together with the suitability of the wood for gun-stocks, has earned it the names of 'Philippine walnut', 'Pacific walnut' and 'Guinea walnut'.

A clearly two-coloured wood, paldao grows in tropical rain forests, where there is no distinction of seasons, so its annual rings are obscure. The pores are large and diffuse, giving vessel lines that appear as coarse scratches on longitudinal surfaces. Paldao is hard and remarkably strong.

The tree that yields this elegant veneer wood grows only in New Guinea and the Philippine Islands, where it is known by the Tagalog name of 'dao' (pal being the Spanish for wood). It grows to a height of 125 feet with girths up to 10 feet, and is strongly buttressed at the base. It is commonest in fertile lowlands beside streams. The bark is smooth and steel-grey, and its surface is regularly shed in scroll-shaped pieces though it is stringy within. Below the grey surface comes a red layer, then a pale pink one, and then a second red layer. The sapwood is thick and pale pink in colour.

The large leaves are placed alternately, but bunched together, at the ends of stout twigs. Each is pinnately compound, with about six pairs of glossy, light green, smooth leaflets. A single leaf is about one foot long by four inches wide. Small flowers are borne in short spikes, and are followed by hard round seed-pods.

In the Philippines paldao is used extensively for house-building and furniture. Large logs are hauled in from the swampy jungles by teams of carabaos or water-buffaloes, or their modern equivalent, the crawler tractor. The brilliant coloration is found only in the relatively small heartwood, which has a high value for export.

28 Pearwood *Pyrus communis*

F. Poire, Poirier   G. Birne, Birnbaum, Holzbirne
I. Pero commune   S. Pera, Peral
FAMILY: Rose (Rosaceae)   KEY: D7   SOURCE: Europe, USA

Fig. 50 Simple, long-stalked leaves and oval fruit of the orchard pear, source of pearwood

Pearwood, also called 'fruitwood', is obtained from either the orchard pear or the wild pear tree. It is a lovely rich red in colour, remarkably even in grain, and without any apparent figure. This lack of conspicuous features makes it fairly easy to identify, as is shown in the example on page 82. Sapwood, where present, is pale yellowish-white, and where heartwood is formed there is an abrupt change to the deep red hue.

Pearwood was formerly widely used for mathematical instruments and similar devices where a smooth, hard, reasonably strong and very stable substance was required. Set-squares, T-squares, drawing-boards and rulers were usually constructed by the accurate machining of pearwood. Nowadays it has been replaced for nearly all these purposes by plastics, which can have the great advantage of being transparent.

Pearwood is still used for decorative veneers, because of its attractive and unusual colour. Wood-sculptors like it because it can easily be carved, remains stable, and has a glorious hue that is appropriate to certain subjects. Turners use it for bowls, and it is widely employed to make bread-boards, tableware and other ornamental objects that call for a pleasing, reliable timber of distinctive appearance.

The wild pear is usually a small and slender tree, and large pieces of wood are hard to find. It grows sparingly in European woodlands, and may reach a height of 50 feet, with girths up to 8 feet round, though such trees are exceptional. On the wild form – though not in the orchard varieties – the twigs bear long brown spines. Bark is grey and broken into shallow, squarish plates. The leaves are oval in shape, with a pointed tip and a glossy upper surface, and are remarkable for their long stalks. The flowers are always white – never pinkish as on an apple tree. They open in early May, ahead of the apple blossom, as a lovely shower of white flakes amid the pale green unfolding leaves. The fruit is of course pear-shaped, and has a rounded base, unlike the hollowed-out base of an apple.

Wild pears are hard and woody, with an acid taste. As a rule they only attract the birds, but there is a time, after winter frosts, when it is just possible to nibble them raw. Despite its unpalatable fruit the wild tree is the ancestor of all the garden pears used for cooking, as delicious dessert, for canning in syrup, or for making the strong alcoholic drink known as perry.

29 Pine, Ponderosa *Pinus ponderosa*

F. Pin ponderosa   G. Gelbkiefer   I. Pino ponderoso
S. Pino ponderoso
FAMILY: Pine (Pinaceae)   KEY: B6
SOURCE: Oregon and neighbouring western States

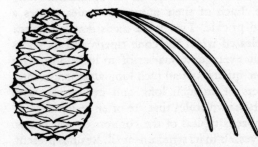

Fig. 51 Ponderosa pine bears slender needles in threes, and has a huge cone with prickly scales

Ponderosa pine, also known as 'western yellow pine', flourishes as a wild tree in all the western States. Its range extends from British Columbia southwards through California to Mexico, while inland it is found as far east as Nebraska. Like all true pine trees, it bears its needles in clusters, and it is one of the few species to have *three* needles per cluster. Most common species, such as lodgepole pine, bear needles in pairs, while a few, including the eastern white pine, have five needles in every cluster. Ponderosa pine grows in the drier zones among the Rocky Mountains and neighbouring ranges, and is able to thrive under lower rainfalls than most conifers. It has been introduced to several countries overseas, and plantations have been formed in South Africa,

143

Australia and New Zealand. In Britain it is occasionally seen as an ornamental tree, and a few trial plots have been established to test its productivity under a cooler and moister climate.

Ponderosa pine can attain impressive sizes and ages. Trees have been measured 232 feet tall and 30 feet round, and an actual count of annual rings, on a tree felled in eastern Oregon, revealed an age of 726 years. Fast growth of a useful timber, on relatively poor soils under difficult climates, makes ponderosa pine a highly valued timber tree in the West. It springs up readily from self-sown seed, and is easy to raise in nurseries and establish by planting.

Typically, ponderosa pine makes an erect tree with a columnar trunk of good timber form, and a rather light crown. Its bark is yellowish or dark reddish brown, breaking into irregular scaly plates that become large and thick in old trees. The needles, which are always in threes and last for three seasons, are remarkably long, averaging six inches, dull green in colour, and tough. Male flowers, which open in spring, consist of clusters of yellow anthers, which scatter pollen and then wither. The female flowers or conelets are greenish structures, one inch long, and take eighteen months to ripen into woody cones. Cones are oval in outline, bright brown in colour, and measure about six inches long by three inches wide. Each of their numerous scales carries a strong, sharp, reflexed prickle. Two large seeds develop below each scale, and are released when the cone ripens and its scales expand. The seed-grain averages a quarter of an inch long, with a thin wing about three-quarters of an inch long attached. Winter buds are three-quarters of an inch long and cylindrical, with closely pressed reddish-brown scales that are often resinous.

Ponderosa pine timber is typical of the conifers or softwoods. There are no pores or vessels in its structure at all. Resin is present, being held in ducts or resin canals that run all through the wood. It is revealed by a strong smell of turpentine from all freshly cut surfaces, and by crystalline blobs of amber rosin that form if a living tree is injured. There is a sharp contrast between the red-brown summerwood band and the wider, yellow springwood that together make up each annual ring. The timber as a whole is light and fairly soft. It is easy to work yet strong enough for a very wide range of uses. Sapwood, where present, is creamy white and well defined. Despite its resinous character ponderosa pine is not naturally durable, and needs treatment for outdoor use.

A constant feature of all true pine timbers of the genus *Pinus*, which includes ponderosa pine, is that the knots are always in distinct groups, with knot-free timber between each group. This results from the branching patterns, for in pine the side branches always grow out from set points, called 'nodes', and usually only one node is formed each year. Other conifers also bear some branches separately, between their nodes, so the presence of inter-nodal knots proves that an unknown timber *cannot* be a pine.

Ponderosa pine is a leading commercial softwood in the West, available in large quantities and adaptable to many purposes. It is employed in house-building, heavy constructional work, the indoor finish of houses, cupboards, doors, flooring, joists and rafters. Out of doors, preservative-treated pine is used for fencing, telegraph poles and railroad ties, also as pit props in mining. Much is used for general carpentry, and for making wooden boxes and packing cases. Despite its resinous nature, it can be used for paper-pulp, and in the manufacture of all sorts of 'man-made board' including chipboard, hardboard and insulation board. It is also a satisfactory firewood.

Though widely employed in many everyday jobs, ponderosa pine is an attractive timber, and selected logs are frequently cut into decorative veneer. This is usually peeled by the rotary process to reveal the flamy pattern of the distinct grain, and to expose the grouped knots as clear circles. The general colour of such veneer is clear soft yellow, figured with reddish-brown along the summer-wood bands, and punctuated by the dark rust-red to blackish-red roundels of the knots. This creates an authentic impression of Western, ranch-style woodwork, since ponderosa pine was the usual constructional softwood throughout the western States during pioneer days – as it is today.

30 Primavera *Tabebuia donnellsmithii*

FAMILY: Bignonia (Bignoniaceae)   KEY: B7
SOURCE: Central America

Fig. 52 Primavera bears palmately compound leaves and large, yellow, trumpet-shaped blossoms

The beautiful Spanish name of this Central American timber has the primary meaning of 'spring-time', but it can also mean 'beautifully coloured thing', or 'figured silk'. Both descriptions fit this lovely timber, which has also been called 'sunray' in English-speaking lands. It is cut from a tall tree that grows along the western slopes of the Central American ranges, in southern Mexico, Guatemala, Salvador and Honduras. This tree has a white bark and large leaves compounded of several leaflets spreading out like the fingers of a hand. During the dry season the tree stands leafless, and soon after it bears magnificent clusters of large, yellow, trumpet-shaped flowers, standing out against the sky. These are followed by pod-like fruits, holding several seeds.

Primavera timber is used locally for everyday work, including the siding of houses, planking for boatbuilding, furniture and plywood. Selected logs showing a lustrous ribbon figure on cut surfaces are exported to North America and Europe for ornamental veneering, or for use in the finest cabinet-work. It is handled in the same way as satinwood, though its good carving properties and stability have earned it the name of 'white mahogany'.

31 Purpleheart *Peltogyne porphyrocardia*

F. Bois d'amarante, Bois pourpre   G. Amarant, Violettholz
I. Amaranto violetto   S. Palo morado, Amaranto
FAMILY: Sweet pea (Leguminoseae)   KEY: C2
SOURCE: Central and South America

Fig. 53 Purpleheart leaves consist of two leaflets each; the fruit-pod is one-seeded

The aptly named purpleheart cannot be mistaken for any other timber, for its whole surface is a bright, clear, purple colour. This is due to a remarkable natural pigment, found only in the heartwood; the sapwood is whitish, with purple streaks, and remains so. When the heartwood is first exposed by cutting, it is horn-coloured. Exposure to the air causes it to turn, within a few days, bright purple on the surface. But this colour is only about a twentieth of an inch deep, and further cutting will expose pale surfaces, which become purple in their turn. Long exposure to sun and rain renders the surface black, but it is constant enough when used indoors.

Purpleheart is remarkably heavy, strong and tough, weighing 54 lb. to the cubic foot. It grows in Central American rain forests and shows no marked features of annual rings, rays, pores or grain. Figured wood is rare and correspondingly valuable. Local uses for material not found good enough for export include wheel spokes, house-framing and heavy constructional jobs. In America and Europe small quantities of purpleheart are used for fine turning and cabinet-work, but the main demand is for veneers and inlays.

Purpleheart is also known as 'amaranth' or 'purplewood'. A Portuguese name common in Brazil is 'pau roxo', while in Panama it bears the Spanish names of 'palo nazareno' and 'palo morado'. An expressive Indian name used in Brazil is 'coataquicana', meaning 'monkey's hammock', because of its long, slender, flowering branchlets. In Guyana it is called 'sakavalli' or 'kooroobooelli'.

Purpleheart is cut from several species of tall trees that belong to the genus *Peltogyne*, the commonest being *P. porphyrocardia*. They grow along riverbanks and lake-shores in Central and South America, from Panama to Venezuela, Trinidad, Surinam, Guyana and northern Brazil. Some achieve heights of 125 feet and girths of 12 feet, with tall boles, clear of branches, 40 feet long. The leaves are leathery in texture and each consists of only one pair of pointed, oval leaflets – a rare arrangement. The small white blossoms, borne in clusters, are followed by one-seeded fruit-pods. The bark is smooth and grey, and the trunk is round, without buttresses.

32 Rosewood, Brazilian *Dalbergia nigra*

F. Palissandre du Brésil   G. Palisander, Rio Palisander, Rosenholz
I. Palissandro del Brasile   S. Palo de rosa, Palisandro de Rio
FAMILY: Sweet pea (Leguminoseae)   KEY: E8   SOURCE: Brazil

Fig. 54   Brazilian rosewood
has a compound leaf
and flat seed-pods

This beautiful and costly wood, also called 'Rio rosewood', 'Bahia rosewood' and 'palisander', is cut from a slender tree that grows in eastern Brazil, around Rio de Janeiro. Though it occasionally reached 125 feet high, large examples are now scarce, and the diameter of the logs available in commerce is limited further by the removal of the greyish sapwood. Rosewood has a compound leaf, built up of about seven oddly shaped leaflets. Its yellow flowers resemble those of a sweet pea, and it bears hard seeds a few at a time, in flat pods.

Rosewood draws its name from a fragrant oil that pervades it – with the scent of rose blossom. Its value rests in several properties of exceptional weight, smoothness, strength and hardness, united with attractive appearance. The ground colour is a warm reddish brown, enlivened with irregular patches of paler golden brown and others of a violet-brown hue.

A network of black streaks or veins runs throughout all these colours. Rosewood is difficult to work, but can be brought to a fine polish.

Brazilian rosewood was for long a leading timber for luxury woodwork, including pianos, billiard-tables, cabinets, dressing-tables and fine chairs. Many choice pieces have been made during the 350 years that this rare wood has featured in commerce, and are rightly prized. Today scarcity and cost limit the regular use of rosewood, in solid form, to small objects such as cutlery handles, brush-backs, trays, artistic turned goods, wood sculptures and the choicest of furniture. As veneer, it is widely employed for radio cabinets, cocktail cabinets, coffee-tables and similar ornamental purposes where the high cost of a decorative finish is immaterial. It is also used in marquetry and inlaid-work.

In Brazil this timber is known as 'cabiuna', 'jacaranda', 'pau rosa' or 'pau preto'. It is one of several species of *Dalbergia* found in South America, Central America and India, all of them yielding ornamental rosewoods. A characteristic feature of rosewoods is a gum that fills the rather coarse pores. When these are seen as vessel lines on longitudinal surfaces, by diffuse light, this gum appears black in this species, though reddish, yellow or white in others. But in all cases it reflects direct light, giving the surface a fascinating silvery sparkle; tilting the specimen will reveal this.

33 Sapele *Entandophragma cylindricum*

FAMILY: Neem tree (Meliaceae)   KEY: D8   SOURCE: West Africa

Fig. 55   Sapele bears compound leaves and oblong seed-pods holding winged seeds

Sapele, or 'sapele mahogany' as it is often termed, is one of several African timbers that are harvested and marketed as substitutes for the true Central American or Honduras mahogany (see p. 129). Since it belongs to the same botanical family, the Meliaceae, it has broadly similar properties as well as likenesses of colour, grain and appearance.

The main points that mark out sapele from African and Honduras mahoganies are its clearer annual rings, its greater hardness and weight, and its cedar-like smell. In colour it is rather brighter pinkish red; sapwood, where present, is pale pink. The distinct pores are clustered rather than being evenly spread through the wood.

A remarkable feature of sapele is that the grain is interlocked, and changes in its general direction occur at frequent, though irregular, intervals. When the log is cut radially, this peculiar grain arrangement shows clearly as alternating light and dark stripes. 'Striped sapele' is very popular as a decorative surface veneer for high-grade furniture, particularly book-cases and cabinets for all purposes, from cocktails to radio. Most sapele is used for veneer; the interlocked grain makes it less suitable than mahogany for solid construction, as it is apt to warp and usually 'works' unevenly under cutting tools.

Sapele grows throughout the West African rain forest as a tall, sturdy tree with a buttressed base. It bears compound leaves, small clustered yellowish-green flowers and oblong seed-pods holding winged seeds. The bark is thick but smooth. Local names are 'aboudikro' and 'sapelli', while sapele is the common trade name throughout the world.

34 Satinwood, East Indian *Chloroxylon swietenia*

F. Citronnier de Ceylan   G. Ostindisches Satinholz
I. Satinato, Citrino Indiano   S. Satinato de Ceilan
FAMILY: Rue (Rutaceae)   KEY: B8   SOURCE: Ceylon

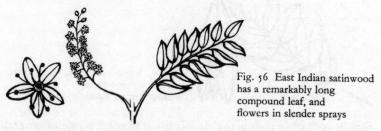

Fig. 56 East Indian satinwood has a remarkably long compound leaf, and flowers in slender sprays

The satinwood tree is native to Ceylon and South India, where it is said to reach a height of 80 feet and a girth of 9 feet. But all the accessible full-sized trees have been felled for their valuable timber, and supplies today come from smaller specimens, rarely more than 50 feet tall and 4 feet round. The tree is easily known by its remarkable compound leaf, which grows from one to two feet long and consists of from ten to twenty pairs of leaflets, each about one inch long. One side of each leaflet-blade is wider than the other; there is no terminal leaflet. Twigs, leaf-stalks and flowers are all clad in short grey hairs. The small flowers are creamy white in colour and grow in clusters near the tips of the twigs.

Satinwood comes from regions with long, dry seasons and matures slowly. It is very heavy, weighing 55 lb. to the cubic foot even when seasoned, hard, tough and naturally durable. Heartwood and sapwood have the same yellowish colour. Annual rings and rays are obscure, and pores are diffuse and very fine. The value of satinwood lies in the wonderful surface lustre that gives it its name. This is combined with beautiful broken stripe and mottle figures due primarily to its irregular grain. It has long been used in the solid form by Sinhalese, Indian and Western cabinet-makers and wood-turners. Today most of it is harvested for radially sliced veneers, used on fine furniture, often in contrast with darker-coloured woods.

35 Sycamore Maple *Acer pseudoplatanus*

F. Sycomore, Erable sycomore   G. Ahorn, Bergahorn
I. Accro montano   S. Sicómoro
FAMILY: Maple (Acceraceae)   KEY: A6
SOURCE: England, Europe, USA (introduced)

Fig. 57 In sycamore maple the lobes of the leaf are rounded; winged seeds face each other at a shallow angle

Sycamore maple is the largest of several kinds of maple that are native to Europe. It is occasionally planted as a shade tree in America, and its timber is imported either under its true name or else, after dyeing to a pleasing grey shade, as 'harewood'. Sycamore maple grows on all the mountains of Central and Southern Europe, including the Alps and the Apennines. It is not truly native to England, but was introduced in the Middle Ages and has long been naturalized. It springs up in woods and waste places everywhere from its plentiful winged seeds.

The sycamore maple tree is best known by its leaves, which have *rounded* lobes, lacking sharp points. Its bark is a distinctive metallic grey, with pinkish tints in the under layers, and on old trunks it repeatedly breaks away in shallow, circular platelets or discs. The flowers open late, about June, and hang down from the branches in loose clusters; they are pretty but not conspicuous, since they are greenish yellow in colour. Sycamore maple yields little sugar sap, so it is never tapped.

Sycamore maple timber resembles white maple in its general character, but is distinctly softer, and shows a greyer shade of white, rather than a brownish one. Many of its uses are the same, but it is seldom employed for flooring because it is less hard.

Conversely, it is used more often for carving, since it is easier to work. An example of fine carving in sycamore is the 'butter print', a wooden mould designed to shape pats of butter into figures of flowers or animals, which was popular on old-fashioned dairy-farms. Because of its clean white appearance and smooth finish, sycamore is widely used in Europe for turned bowls, platters and tableware. It is also made into rollers for textile machinery because it never stains the cloth. Much good modern furniture is also made of sycamore maple wood.

By long tradition certain parts of violins, fiddles, guitars and similar stringed instruments are always made from sycamore maple. These are the sides, the back and the stock, along which the strings are stretched. But the front, or belly, must never be made of sycamore because it lacks natural resonance. The timber used for the belly is always a kind of spruce called 'Norway spruce' though most of it comes from the Alps! The number of rings in each inch of width governs the timbre of the instrument, and slow-grown trees suitable for violin bellies are rare except at high altitudes. Many violins show on their underside a handsome dappled effect of rippling light and shade. This is known as fiddle-back figure, and the sycamore maples that yield it are very valuable.

Sycamore maples that are believed to hold wavy grain or fiddle-back figure are usually reserved for slicing into veneer. Very high prices, up to £2,000 ($5,000) for one tree, have been paid for large, old, open-grown trees in Yorkshire. Some experts claim to detect the presence of wavy grain by cutting a 'window' in the bark of the tree, in order to see the wood beneath it. Others claim to recognize it from the character of the bark! This tree can grow very large, records for England being 117 feet in height and 20 feet in girth.

The base of an old sycamore maple is usually fluted or buttressed to some degree, and this brings in another element of variation in grain that is used to good effect by skilful veneer-cutters. The pattern can also be enhanced by 'weathering' to make the faint coloration darker. Originally this implied long exposure to sun and rain, but nowadays it is speeded by chemicals. The whiteness of sycamore makes it very suitable for staining, and it is frequently coloured grey to make 'harewood'.

36 Teak *Tectona grandis*

FAMILY: Verbena (Verbenaceae)   KEY: E9
SOURCE: Burma, Southern India, South-east Asia

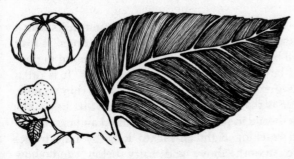

Fig. 58 Teak has a large simple leaf, and a woody fruit; at left, a seedling sprouts from a kidney-shaped seed

Long before the first European navigators had reached India, Eastern people had realized the unique properties of teak, which grows in southern India, Burma, Thailand and Java. Teak timber had been used for the building of Arab dhows and Chinese junks, and Portuguese, Dutch and British seamen soon applied it to the repair of their oaken craft, damaged by the seas on the long voyage round the Cape of Good Hope. It was found to be remarkably strong and naturally durable, resistant to both insect attack and fungal decay, and even to termites and marine borers. A first-class wood for house-building, it is also well adapted to heavy construction such as bridge-works, dock gates and railway engineering and to fittings needing strength and resistance to chemical attack. It is used, for example, for vats and as laboratory benches, since it is not harmed by acids or alkalis. It makes sturdy long-lasting garden furniture that will endure out of doors in all weathers. Modern designers, particularly in Scandinavia, have adapted it to artistic indoor tables, chairs and cupboards too. Much is used for ships' decks, and in boatbuilding generally.

In appearance teak is an even mid-brown in colour, with a hint of olive-green or gold. It has a dull, rough surface, without lustre, a curious leathery smell and an oily feel. The annual rings are clearly marked by the grouping of large pores in circles, following the start of each year's growth. This gives a definite grain, although the wood is one-coloured and its rays are obscure. Teak holds a natural oil, but benefits from oiling after long exposure to the weather. This restores its natural colour, which otherwise becomes bleached white by the sun. Sapwood, where present, is pale yellow, and even the heartwood is yellow rather than brown when freshly cut. Teak is so hard that ordinary nails cannot be driven into it, and it is therefore pre-bored, or fastened with wooden pegs or screws. In contrast to oak, it does not corrode iron fittings. It is exceptionally heavy, and freshly felled logs sink in water.

Teak grows as a magnificent tree in monsoon rain forests with a marked dry season, as revealed by its clear annual rings. It may reach 150 feet tall by 40 feet round, but it is seldom found in pure stands and must be sought out in mixed woodlands. It is often first noted by its greyish bark, which shreds off in thin strands, and a further clue is that there are no buttresses at its base. The leaves, which are borne in pairs, are oval, pointed and exceptionally large, often reaching eighteen inches long by nine inches across; they are smooth above and hairy below, and show prominent curving veins.

Small flowers, borne in upright clusters, are followed by large fruits, three-quarters of an inch across, with a wrinkled outer surface. Each holds four large, hard, kidney-shaped seeds. Seedlings are easy to raise, and extensive plantations of teak have been made in India and Indonesia, to ensure future supplies more profitably than those from wild trees scattered through remote jungles.

Most teak however, is still harvested from naturally grown trees. Forests are divided into large sections, each of which is explored in turn to maintain continuous supplies. When mature trees big enough for felling are found, they are hammer-marked by the timber concern to establish ownership, and are then ring-barked. Ring-barking involves removing a strip of bark and sapwood all round the tree. This kills it by denying all nourishment to its roots, since nothing can descend from the tree's crown of foliage. But the tree takes a whole year to die, and meanwhile the leaves transpire all the sap stored in the trunk, before they finally wither. This makes the log light enough to transport by floating.

After felling and cross-cutting, the huge logs or baulks of timber are hauled to the nearest waterway by elephants. These huge powerful beasts have been captured from wild herds and trained for haulage work. They will also, when instructed, lift and stack logs with the aid of their great trunks. The timber is floated for great distances downstream along broad rivers like the Irrawaddy and then landed at the seaports for shipment to North America or Europe. True teak is so called in all European languages. In Burmese it is 'kyun', and in Hindustani, 'sagun'. Other so-called 'teak' woods from Africa or elsewhere are not botanically related, and are named solely for similar appearance and working properties.

37 Walnut, Australian *Endiandra palmerstonii*

F. Noyer d'Australie   G. Australischer Nussbaum
I. Noce Endiandra   S. Nogal Australiano
FAMILY: Laurel (Lauraceae)   KEY: E10   SOURCE: Queensland

Fig. 59   Australian walnut has a simple, leathery leaf and a nut-shaped fruit

The popularity of Circassian or English walnut, and of American walnut (both described later), has led to a world-wide search for similar timbers. Australian walnut, also known as 'Queensland walnut', 'orientalwood', or 'Australian laurel', is one of the most suitable of these. It is not a true walnut botanically, since it belongs to the laurel family and bears irregular, oblong, leathery evergreen leaves, each being about three inches long, rather like a laurel bush. Its fruit, however, is walnut-like, being a globe about two and a half inches across, bluish black in colour. The flowers are very small, and are borne in open clusters.

Australian walnut grows in the sub-tropical forests along the coast of Queensland, as a tall tree up to 140 feet high, with a clear bole up to 80 feet, and a diameter of 10 feet. It has a pale brown sapwood, but this is seldom seen on imported timber. Freshly cut wood has an objectionable smell, which vanishes later.

The heartwood is a beautiful soft brown, suffused with pinkish overtones. Blackish-brown or even black streaks run lengthwise and mark the outer edges of the annual rings. It is heavier than English walnut, and can be distinguished from it by its greater range of colouring and less distinct annual rings. It is not so easy to work, but can be made to serve most of the purposes to which the northern walnuts are applied. It yields highly decorative veneer, and is popular as a surface finish for pianos, high-class furniture and cabinets. Some is used for joinery and panelling, and it proves a good turnery wood.

38 Walnut, Black American *Juglans nigra*

F. Noyer Americain   G. Schwartznüss, Amerikanischer Nüssbaum
I. Noce nero d'America   S. Nogal negro Americano
FAMILY: Walnut (Juglandaceae)   KEY: E11   SOURCE: Eastern USA

Fig. 60   Black American walnut bears a compound leaf with many pointed leaflets, and has a rugged nut

Black American walnut grows wild as a tall tree of the rich bottomlands in natural woodlands right across the central eastern and mid-western States. It is not found, except under cultivation, in Florida, the western States or northern New England, though its range just brings in southern Ontario. Cultivated strains are grown for fruit or ornament throughout North America, and also, occasionally, in Europe. It can form a very tall tree, up to 150 feet high by 20 feet round, and stand for over 250 years.

Black American walnut is so called to distinguish it from Circassian or 'English' walnut (see p. 157). Differences begin with the bark, which is dark grey to black in American walnut, rugged and split into squares – contrasting with the pale grey, ribbed pattern of English walnut. American walnut has the larger leaf, often over a foot long, made up of about ten pairs of leaflets, against about five pairs in the English tree; these leaflets are more slender ovals, and end in long points. In the wild tree, the nuts have a thicker, rougher shell and are less pleasantly flavoured than those of English walnut; but in cultivated strains, increased by grafting, the American nut-kernels are of excellent form and flavour.

Black American walnut has a very distinctive timber. The thin sapwood is pale yellow, and the heartwood within it is a rich, deep chocolate-brown. In some examples it is dark, or even deep purplish-brown to black; any paler wood shows blackish streaks. It is easily distinguished from the much paler, 'milk-chocolate' brown of English or Circassian walnut, though the figuring occasionally shows resemblance. American walnut has a curious odour and even a taste, mild yet distinctive; hence its old name of 'gumwood'.

The excellent physical properties of American walnut were quickly appreciated by the early colonists, who used it for furniture, gun-stocks and other exacting work needing strength and stability. The texture is coarse but uniform, and the heartwood is naturally durable. The remarkable colour of this wood has ensured its continued use in fine cabinet-work and also as decorative veneer. Its grain is usually straight, but wavy or curly examples prove highly decorative. Selected pieces show luxurious blends of deep violet-brown colours that no other timber can match.

39 Walnut, Circassian *Juglans regia*

F. Noyer   G. Welchnüss, Walnüss, Nüssbaum   I. Noce   S. Nogal
FAMILY: Walnut (Juglandaceae)   KEY: E12
SOURCE: Turkey, South-east Europe, USA (introduced), England (introduced)

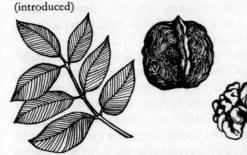

Fig. 61 Circassian walnut leaves have few leaflets, with rounded outlines; on right the fruit, or walnut, with kernel

Circassian walnut, also called 'Persian', 'European' or 'English' walnut, originates in Turkey and neighbouring regions of Asia Minor. The finest coloured and figured wood is obtained from the province of Circassia, near the Black Sea, south of the Caucasus mountain range.

The Romans discovered this remarkable tree on their conquering expeditions to the East, and called it *jovis glans*, meaning 'the nut of the great god Jove [or Jupiter]'. From this we get the scientific name *Juglans*. The specific name *regia*, added by the Swedish botanist Linnaeus, means 'fit for a king'. Both are apt names for a tree that yields a deliciously flavoured and nutritious nut, as well as a remarkably strong and attractive timber.

Roman colonists and monastic gardeners carried this desirable, easily propagated tree to western Europe, where it gained several local names. It is walnut or 'welshnut' in England, 'walnoot' in

Holland, and 'Welchnüss' or 'Walnüss' in Germany, from the Teutonic root 'wal' or 'wealh', meaning 'foreign'. In Ireland it is 'gallcno', and in Wales, 'cneuen ffrengig', meaning 'French nut', while in France itself it is known simply as 'noyer', the nut tree, and its fruit as 'noix', the nut of all nuts. Walnut was one of the first trees introduced to New England by early colonists. It is now grown right across the United States, also in New Zealand, South America and Australia.

Walnut comes into leaf very late, about mid-May, and the leaves have a characteristic coppery-brown shade as they slowly unfold. Each leaf is about four inches long and compound, made up of some nine oval leaflets with rounded ends. When crushed, they have a rich aroma. Their juice stains the fingers brown – gipsies once used it as a sun-tan lotion. These features, and the fact that leaves and buds are set singly on the twigs, distinguish walnut from ash, which bears similar compound leaves, set in pairs.

Circassian walnut bark is grey, like that of ash, but has a bolder, less regular pattern of ribs with a silvery-grey metallic sheen. Black American walnut has a very different bark – darker grey and broken into roughish squares.

Male and female catkins open at the same time as the leaves, all on the same tree. Male catkins look like fat greenish caterpillars, and fall as soon as the pollen has been scattered on the winds from their numerous blossoms. Female catkins, in groups of two or three, are each shaped like an Italian wine-flask, with two horns or stigmas at the tip. They develop rapidly into a green plum-like fruit, which is often picked at the half-ripe stage for pickling in vinegar. The green outer pulp soon withers and the hard-shelled, crinkly nut inside it is revealed. The tasty and nutritious kernel within it is divided into curious lobes, and an odd papery membrane runs across the heart of the shell, almost, but not quite, splitting the kernel. This kernel is really a pair of seed-leaves modified to make storage organs. If you sow a walnut, it will send out a sturdy root and a shoot that bears typical compound leaves.

Walnuts are widely grown for their nut crop in California, South Australia, France, Italy, Turkey and Greece. Selected strains, which bear heavy crops of large nuts, are propagated by grafting in orchards.

Circassian walnut is never a very tall tree – 80 feet is the maximum, but it can attain great girth and age, and stems are known that are 22 feet round and 400 years old. A curious feature of its twigs is that their stout pith is hollow except for small membranes that cross it at intervals – what the botanists term 'laminated pith'; a slanting cut reveals this key feature. The leaf-scars are large and prominent, with 'horse-shoe nail' markings; the winter buds are greyish-brown furry buds.

The Circassian walnut tree has a thin outer zone of pale yellow sapwood, rarely used commercially. Within this lies the magnificent heartwood, hard, strong, heavy and naturally durable. Its general colour is a warm, greyish-brown, with a hint of chocolate, and this is suffused with a pinkish hue, more apparent in some places than others. A second, distinct pattern of fine black, dark brown or dark grey lines, associated with the outer boundaries of the annual rings, runs throughout this, sometimes blending, sometimes contrasting in outline. Even an ordinary, unselected piece of Circassian walnut shows attractive figure and character, and selected pieces give the most beautiful effects of colour that any wood can show.

Sums of £1,000 ($2,400) or over are occasionally paid in England for large trees. These are felled by cutting below ground-level, since an exceptionally intricate figure is found in the buttressed base where the roots leave the trunk. Walnut veneer is applied to every kind of decorative surface, particularly wardrobes, dressing-tables, television cabinets, and automobile fascias.

Walnut used in solid form usually comes from branches or trunks too small for slicing into veneer. It is widely used for table-ware, including turned bowls, bread-boards, spoons, handles for knives and forks, and toast-racks. Wood-sculptors employ it widely for carvings with a lustrous surface that are durable and stable in shape. Walnut is by far the best wood for gun-stocks, again because of its exceptional stability when seasoned. Once a stock has been carved to the user's needs, it will never shrink or warp despite years of tough use. The dense nature of walnut also enables it to take the recoil of the shot without distortion.

40 Zebrawood *Microberlinia brazzavillensis*

FAMILY: Sweet pea (Leguminosae)　　KEY: B9　　SOURCE: West Africa

This very striking timber is known at once by its bright yellow bands alternating with dull brown to blackish strips, giving a zebra-stripe appearance. These bands follow the general run of the annual rings, right down the length of each log, but do not coincide with them. The rings can, however, be picked out, particularly on the end-grain. Sapwood, which is rarely seen on imported material, is pale yellow in colour and shows no stripes.

Zebrawood grows in Cameroun and neighbouring States of West Africa, on and around the Equator. It is a small tree, rarely taller than 65 feet or stouter than 8 feet round. The leaf is compound, with about eight pairs of leaflets. The flower resembles that of a sweet pea, but is larger and orange coloured. Seed ripens in broad flat pods.

The brilliant coloration of zebrawood presents a challenge to the designer of fine joinery and cabinet-work, for it must be used skilfully to secure an artistic, but not garish, effect. It is popular both in America and on the European continent for shop- and restaurant-fittings, and large panels in furniture; for these purposes it is usually employed as veneer. Decorative turners and carvers also employ it for exceptional effects.

Zebrawood is one of several timbers that are called by this or similar names, such as 'zebrano', in various parts of the world. It is also known as 'zingana', especially in America and Germany.

Fig. 62   Compound leaf
and seed-pod
of zebrawood

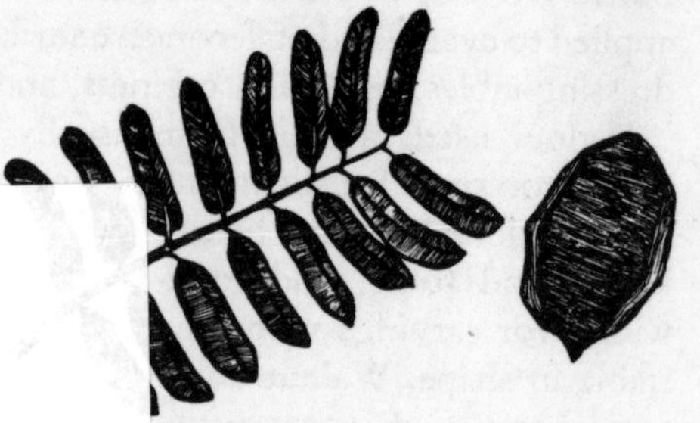

of the scientific staffs of the following
literature and botanical specimens:
culture, Washington, D.C.; Forestry
nent, Ottawa, Canada; Commonwealth
Botany, Oxford University; Forest
f Technology, Princes Risborough,
Edinburgh; and Forestry Commission

The frontispiece and the forty pictures of tree features are from original drawings by Miss Anne Semple.

The identification keys are based on those in Dr Alfred Schwankl's book *Welches Holz ist Das?*, published by Franckh'sche Verlagshandlung, Stuttgart, whose permission to reproduce this material is gratefully acknowledged. Illustrations on the first seventy pages of the text are taken from the same source.

The wood specimens were selected from the stock of Art Veneers Company Ltd, Mildenhall, Suffolk, England, who for many years have specialized in providing veneers to all parts of the world.